A BOOK of BALANCE

A Book of Balance

Kogi Wisdom
for a
Good Life
and
Thriving Earth

LUCAS BUCHHOLZ

Translated by Fabienne Balmer

Edited and Adapted by Jackson Gore

HarperOne
An Imprint of HarperCollins*Publishers*

FIRST EDITION

Designed by Janet Evans-Scanlon
Canvas texture used throughout © Mehmet Gokhan Bayhan/stock.adobe.com
Leaf ornament used throughout © Tally 18/stock.adobe.com
Photograph on p. i © Lukasz Malusecki/Shutterstock
Photograph on p. 2–3 courtesy of the author

Library of Congress Cataloging-in-Publication Data has been applied for.

ISBN 978-0-06-332990-4

24 25 26 27 28 LBC 5 4 3 2 1

Contents

Before the Journey

The invitation you have received is not from me, but from the Kogi themselves. They have asked you to be here, in this moment, open and receptive to the enormity of the challenges we face as humans, to sit by the fire, and in the vast darkness of potential, reach within yourself to find, and become, the solution.

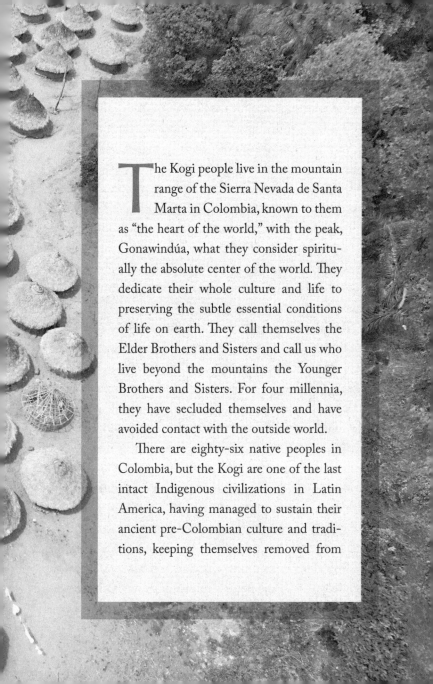

The Kogi people live in the mountain range of the Sierra Nevada de Santa Marta in Colombia, known to them as "the heart of the world," with the peak, Gonawindúa, what they consider spiritually the absolute center of the world. They dedicate their whole culture and life to preserving the subtle essential conditions of life on earth. They call themselves the Elder Brothers and Sisters and call us who live beyond the mountains the Younger Brothers and Sisters. For four millennia, they have secluded themselves and have avoided contact with the outside world.

There are eighty-six native peoples in Colombia, but the Kogi are one of the last intact Indigenous civilizations in Latin America, having managed to sustain their ancient pre-Colombian culture and traditions, keeping themselves removed from globalization and modern Colombian society. What the Kogi offer us is an unbroken memory of their culture; it is the original thoughts of their people, their ancestors, the Tayrona.

Very few Kogi have ever left their mountains, let alone their country. But Mama José Gabriel, who was eighty-five years old, had come to Frankfurt, Germany, to speak about the Kogi coffee project that was set up to support the Kogi financially so they could buy back their ancestral land from the Colombian farmers. He had trained to be a representative of his people since childhood, to share a stark message and warning about what wisdom might be lost if the people beyond the mountains did not pursue greater balance.

The Mamos in general are the Kogi teachers of wisdom and the keepers of knowledge. They preserve all that is essential and sacred for the daily lives of their people to be in balance, and they protect whatever falls under their responsibility for life on earth to continue. Their view is focused on a planetary context as much as on their locality. This includes gardening, health, diet, village planning and building work, forest management, marriage counseling, raising children, psychotherapy, music, dance, astronomy, philosophy, and the knowledge of the sacred places and what is required to refine and maintain both inner and external order. The Mamos and their female counterparts, the Sakas,[1] are the spiritual authorities of the Kogi. Their relaxed authority is anchored in their profound knowledge of life and not in a misuse of power. Both the Mamos and the Sakas guide through their actions and by being deeply rooted in their wisdom. Some of them live as advisors in the villages, others live in hermitages in sacred places in the Sierra, where they keep the world in balance through their thoughts and prayers. The Mamos based in the villages accompany the nightly conversations in the *nuhué*,[2] and in offering their advice they help to organize the thoughts of the people. They are also responsible for the four great rituals that take place in the life of a Kogi: welcoming a newborn, the transition from adolescent to adult, marriage, and death.

As I speak Spanish, I offered to be Mama José Gabriel's German translator for the talks and also offered him a place to stay, as the event organizer could not find suitable accommodation. So the Kogi Mamo stayed with me, and what ensued were some of the most hilarious intercultural misunderstandings, such as his dislike of my vegetarian food and request for me to cook him a big rat for dinner. Clearly that was a rather challenging request; however, the next morning I cooked him a traditional German bratwurst, which he thoroughly enjoyed. With a huge smile on his face, and as the only one receiving a sausage, he said, "Sezhankua gave me a sausage. I am thinking good thoughts."[3] Later on I learned that the Kogi always say that wild game animals think free and vibrant thoughts, while human-produced animals think thoughts of submission, captivity, and slavery. So, with this view, his original preference made sense.

We traveled together throughout Germany, and as we approached the end of the tour, Mama José Gabriel invited me to come to his village in Colombia—an experience that felt impossible to turn down.

While in the Sierra, Mama José Gabriel asked me to write this book and help spread the urgent Kogi message. He explained he would guide me on what to write so we Younger Brothers and Sisters could begin to understand the message.

In the Sierra I visited different Kogi villages and listened to the different Mamos as they shared their wisdom and knowledge. Throughout my time I was accompanied by a young Kogi man who translated their words into Spanish for me.

Why have the Kogi decided that now is the time to abandon centuries of caution and speak to us?

What exactly has motivated them to deliver their urgent message and warning?

Although the Kogi have lived secluded for many centuries, from high up in their mountain home they have been closely watching our way of life, and they are concerned. They can see clearly the enormous ecological, social, economic, and individual challenges we are all facing. We are in a time of great upheaval and intensifying change, which is resulting in an increasingly volatile and more vulnerable world. The rules and solutions that were applied to life even just ten years ago are no longer applicable in the present time. It is the sheer extent of the damage and our purely reductionist and materialistic approaches to solving problems of all kinds that have prompted the Kogi to now speak to us. They see the reason for the vast majority of our problems in the way we think and in our cosmology, which is the basis of all our actions.

For more than four thousand years the Kogi have succeeded in upholding their original way of life. That is extraordinary when you account for the devastation brought to the rest of Colombia and Latin America. The importance of maintaining an unbroken memory of their culture has allowed not only their wooden ceremonial houses, the nuhué, to stand in the exact same places for over four millennia, but for the wisdom that has nourished and governed their life to continue without any vast decline.

It is important to explain that there are four peoples that inhabit, and protect, the Sierra Nevada de Santa Marta: the Kogi, the Arhuaco, the Wiwa, and the Kankuamo. These four Indigenous peoples are, in many ways, unified and live a shared vision; however, they do differ, as each has its own territory and responsibilities. The ways in which the peoples of the Sierra Nevada have interacted with outsiders and what they have chosen to say have a long history, guided by the leaders of all four peoples. Over the course of this book, what you are reading is mainly the wisdom and perspective of the Kogi, yet in a few instances, for simplicity, I refer to the Kogi when in fact I am sharing the collective wisdom of all four peoples.

This book is much more than a teaching from the Kogi; it is their urgent message to us, the Younger Brothers and Sisters. The Kogi asked me to write this book for them; it is their book and a unique opportunity for the Kogi to speak to us directly. This is rare, because the quotes, conversations, stories, and myths of the Kogi shared in this book take you straight into their world and reveal a sense of how it is to sit with the elders and to listen to their words.[4]

THE KOGI SPEAK MAINLY FROM THEIR OWN FRAME OF REFERence, and as an Indigenous people, largely about nature and its destruction—which they speak of as direct witnesses over millennia.

When I first heard the Kogi speak about the natural devastation in the world, I was immediately confronted by my own internal conflict and prejudice. I had read or heard so often that the biggest extinction of species since the beginning of history was happening right now, that the climate was changing rapidly, and that humans—specifically those of us who live in the modern industrial world—were to blame. I read of widely publicized environmental conferences where governments are unable to find agreement on even the bare minimum to reverse or halt the destruction of nature, and where some very influential countries do not even send delegates.

I find this all deeply worrying and anxiety inducing—yet I am in internal conflict, questioning, how can I, living in a modern city, truly make lasting change?

While I can ride a bicycle, eat organic food, and renounce

the use of plastic bags, am I wholeheartedly prepared to radically alter my entire lifestyle to enable significant and lasting change? And as much as I, and many other people, want change, where and how does one even begin in a society that relies on comfort and conformity over all else?

To what extent are the Kogi's daily lives and realities, and the underlying principles they live by, transferable to my modern way of living?

To get closer to the answer, it is essential to dive into the Kogi's world of thought and to unearth and understand their principles and actions, to better decipher, step by step, the transferable qualities within their daily existence.

It is important to note that the sections in which the Kogi speak are not reworked; the phrasings remain original and authentic and are sometimes very direct.

The Kogi way of speaking is very different from ours, their speeches can often span broad contexts, and their words can act as a gateway to other realms. As they deliberately do not have a written culture, the spoken word is their way to cultivate their memory. Words are precious to the Kogi, and by speaking them, the words are kept alive. The Kogi begin their threads of thought at the origin, with what they call the original thoughts.

You will notice that only the male wisdom carriers of the Kogi, the Mamos, have their say in this book. This is simply because, being a man, I was invited to spend my evenings and nights after dusk in the ceremonial house of the men. In their culture, men are taught by men and women are taught by women. And of course the wise women of the Kogi, the Sakas, are just as knowledgeable and adept as their male counterparts at accessing, upholding, and sharing the ancient traditional wisdom.

The Kogi's invitation to us is not to visit their tropical mountain home in Colombia. They do not want to be visited and ask to be left alone. The invitation is rather to understand their view of the world. They wish this book to reach us, in our exponentially developing yet decaying modern societies, so that we may be inspired to remember our essence and why we are here on this planet.

THE KOGI BELIEVE THAT OUR LIFE IS AFFECTED BY THE QUES- tions we ask ourselves and others; it is their way to inspire balanced thoughts and actions. They do this ritually around the fire at night in the nuhué, the ceremonial hut. But they also pose questions to themselves over the course of the day. During my time with the Kogi, over and over they asked me: Why don't you people in the modern world think more about life? Why are you only preoccupied with yourself and the material world? Does this make you happy and feel good? The Kogi see that it is time for us to pause and consider important questions, because for them the discovery of these answers commits each of us to a better way of life, to slow down and contemplate our inner and outer worlds. They say that real change on earth begins with recognizing within ourselves our innate nature and our individual tasks in the world.

Due to the high regard the Kogi place on upholding what is intrinsically one's own, they have always been reluctant to reveal their way of living. Mama José Gabriel, for example, only started sharing his knowledge after we had told him about ourselves and our way of life. In the beginning, he had always talked about general principles and the broader view of things. Sometimes, however, we did learn details about the specific cultural solutions the Kogi apply to certain challenges. Often, though, the elder Mamo looked at us confused and asked, "Why do you want to know exactly how we do things? What good does it do for you? It's about you finding your own ways again! You know these things too. You just have to remember."

The Kogi's approach to solving conflicts, maintaining preventive health care, and organizing the needs of humans in harmony with those of all other living beings is anything but primitive. The culture of the Kogi has existed for more than four thousand years without any written tradition. Having a complex system made it possible for the Kogi to constantly adapt to new circumstances while also preserving meaningful tradition. The fact that members live in small huts does not mean they are primitive or have no vital knowledge to share with the world. The well-developed and truly reciprocal social structure of the Kogi is very successful, and even if a direct use of the principle of, say, *esuamas*,[5] is not so clear to us at first, it can inspire us to consider our relationship to our ancestral territories.

The Kogi do not have ready-made solutions for global problems; instead, they offer us a view of the world that differs fundamentally from our underlying, and often limiting, assumptions and ideas. They present us with an opportunity to see through a different set of eyes and to detect something new that can guide us out of our destructive and constricted approaches and patterns.

They do not share these words with us so that we simply fight the symptoms of our imbalanced modern society. They do not offer tools, methods, techniques, or goal-oriented achievement models; they offer something far more valuable: refined ancient principles and perspectives that reveal practical wisdom. As we listen to their ancient wisdom, let us remember that the Kogi don't want us to imitate their way of life; they share their thoughts and words for us to begin a deep process of realization and remembrance within ourselves, and within our own habitat and environment. This book is an invitation to sit, and walk, beside me, as I journey through the villages and land of the Kogi and listen to the words of their ancestors and their hearts.

Like all journeys, we must begin at the entrance and threshold, and that place is the nuhué, an awe-inspiring hut, their space of ceremony and prayer, a place where meaning and wisdom are imbued in every action and word that is offered within the walls of this ancient wooden structure. There the Kogi engage in a practice of posing questions around the fire at night. One person asks the question and then after some silent contemplation, various people offer their views.

We ask you to consider these questions in your own life, in your communities, your schools, your workplaces, your relationships.

On entering the nuhué, one enters a darkness—for the Kogi, it is the darkness of the eternal womb of the universe, in which one's own energies, thoughts, and feelings are nurtured. In the nuhué the different aspects within a person become balanced. This happens in the darkness because in the natural world, that is where the stars become visible. For the Kogi, the stars are an absolute expression of cosmic order that light up in the darkness. The perception of time in the nuhué at night is different, like a spiral, and it creates a sense of many worlds existing at the same moment.

CHAPTER ONE

What Did You See Today?

I had long been aware of how our assumptions and our beliefs of the world affect our thinking and our interactions.[1] Yet, this was only in theory; it was not until I met the Kogi that I experienced it in reality: for the first time in my life I found myself in an environment and group of people who hold truly differing views and life experience to my own.

But let's start at the beginning, how I first heard the words of this people. I had gone to Frankfurt to assist in onstage translation for the visiting Kogi elder. It was a last-minute request from the organizer, who did not have a translator. I hoped my Spanish would be good enough, but without hesitation I felt compelled to help. This was the first time I had been onstage to translate. I was nervous, but excited to be there.

The noise and chatter in the large hall fell still as the Mamo began to speak, his soft yet precise words reverberating through the silence:

My name is Mama José Gabriel Alimaco and I come from the Sierra Nevada de Santa Marta. The Sierra Nevada is the heart of the world. There are four peoples living together: the Kogi, the Arhuaco, the Wiwa, and the Kankuamo. We are the guardians of the Sierra Nevada, the heart of the world.

In the beginning of time everything was in darkness and there was nothing but thoughts. There was no earth, no trees, no animals, and no stones. . . . Then Jaba Sé created Sezhankua in Aluna, the world of thoughts.[2]

Jaba Sé was unable to move.

First Sezhankua created the earth in the form of a small black stone. In the beginning the earth was very, very small and was just stone, nothing more.

Then he created the water and then eventually the Sierra Nevada with its animals, trees, mountains, rivers, forests, and its people: first the Kogi, then the Arhuaco, then the Wiwa, and finally the Kankuamo. The four peoples of the Sierra Nevada are the Elder Brothers, they were created first, and Sezhankua gave them the Sierra Nevada to guard.

Then the earth got bigger. She kept growing and eventually became very big. Sezhankua has given a small territory to us Kogi so that we can take care of it. He also gave the Younger Brother a territory so that they could live there and maintain it.[3]

However, at the time when Sezhankua gave all the peoples their tasks, such as what to do, what they shall be responsible for, and how to take care of the earth, at some point around three or four o'clock at night the Younger Brother fell asleep.

Up to this day the Younger Brother has not been able to speak for very long. He gets tired after eight hours at the latest. We Kogi sometimes speak for nine days and nine nights without a break.

Sezhankua had given the Younger Brothers their territory to live in, containing sacred places to take care of. He gave us the Sierra Nevada, a very small territory, for us to guard the heart of the world, but the Younger Brother has forgotten what Sezhankua told him to do.

The Spanish came to Colombia; the Younger Brothers had their own part of the earth to live on, but they still came to us. Many of us were killed. They brought many diseases and many of us died. They wanted our gold and took it. We had no money and no ships, but we had gold in every house. There was gold in every house and it hung in the trees. It was alive, it was a *pagamento* to the earth.[4] The Younger Brothers took it because they didn't understand. Now the gold is dead and sits useless in your museums. It no longer has a task. However, we Elder Brothers said, "Nevertheless, we will live together with the Younger Brothers because they are our brothers and they have come to us."

From that point on the Younger Brothers started to take more and more land. So we climbed high into the mountains and retreated into caves. The Younger Brothers started to live on the land at the foot of the Sierra, but they did not show respect. They disregarded the animals, they disregarded the trees, they disregarded the stones, the rivers and springs, and they disregarded the sacred places and did not care for them and did not nurture them. That has not changed in the past five hundred years.

We have waited. Some time ago, however, we Mamos saw that it cannot go on like this.[5] Sezhankua has given us the heart of the world to take care of, but we asked ourselves, how can we tend to it when we no longer have access to many sacred places? How can we take care of it when the Younger Brothers disregard our sacred places and we ourselves live high up in the mountains and can no longer reach these places?

If we don't take care of the Sierra, the heart of the world, it will die. It will die, and since it is the heart, the whole world will die. From the heart downward the world is already dead, from the heart upward the world is still alive. The Sierra can only live if we get back to our sacred places and if we connect to life through them. When we make a pagamento in the sacred places, we respect the earth. We give something of ourselves for what we receive so that the earth remains in balance. When you buy something in the store, you also pay a lot of money, but you pay nothing for what the earth gives to you. However, if we do not take care of the earth and nourish her, everything will get even more out of balance and she will die and therefore we will also die.

The power, and piercing nature, of the Mamo's words echoed through the hall, and a breathless silence engulfed the room.

FOR MORE THAN FIVE HUNDRED YEARS, SINCE THE SPANISH conquest, the Kogi have only on a few rare occasions reached out to us, and regardless of how little we have heeded their attempts to highlight ongoing devastation, they haven't given up. The Kogi accepted the necessity of being in contact with the outside world for real change to emerge. Even though the vast majority of us Younger Brothers and Sisters believe that our modern world is separate from the world of Indigenous people, the opposite is true. For a real change to take place, dialogue is imperative, and without first listening there can be no dialogue. We must begin to acknowledge that the survival of Indigenous people and their ancestral traditions is inextricably interwoven with our own survival. That without such groups of people upholding and sharing their knowledge and wisdom, what we lose is beyond comprehension; it is the living embodiment and experience of a deeply reciprocal and highly balanced relationship with the earth that sustains us. It is this knowledge that for centuries has been the basis of the dialogue that has been offered to us, and that we have ignored. Why has the essence of life, or in other terms, the animating vital energy of existence or life force, which influences so much of the Kogi's thinking, and that of many other Indigenous peoples, been largely lost in our thinking? We rarely consider this essence of all life.

OLD ROADS LEAD TO NEW PERSPECTIVES

The doors of the train close behind me, and with that, my partner and I leave behind a cold and gray winter morning in Frankfurt, Germany. Shuffling through the packed train to find a seat, we are surrounded by a vibrant mix of people. It is around seven in the morning and we are heading to the airport to fly to Colombia to meet with the Kogi.

The train rattles along, shaking us into a dull alertness under its bright sterile lights and lack of air. The people around us, staring drearily into their smartphones and blankly into space, most of whom are wearing suits, likely on their way to work at one of Frankfurt's many banks or hotels, share one main commonality despite their external differences: an overriding sense of joylessness. I am jolted into a realization: almost no one seems aware of their surroundings, transfixed as they are by the vibration of a WhatsApp message or an internal monologue.

As quick as this thought comes up, I too am reminded to be present; we have to change trains. The train screeches to a halt, and we disembark. On the platform people are darting in all directions, it is a rush hour melee. It seems as though the morning journey to work is just another mechanical action that takes us from one place to another; it feels totally devoid of vitality or connection. We finally arrive at the airport; relief fills my body, not only from what I am leaving behind, but also from the realization that we are one step closer to what is waiting for us.

Standing in line for check-in, my thoughts begin to wander to the past, and memories start to appear. During my university studies I did internships abroad and stayed in Mozambique, Jordan, Pakistan, Israel, and Palestine. I got to live with, and to some extent know, other cultures. I learned parts of other languages. All these experiences were essential, but not for the reasons I assumed at the time. What I came to witness and comprehend was the astounding depth and far reach of how homogenization had taken over the minds of the people, in both their culture and their way of thinking.

The picture is still very present in my mind: driving from the Aeroporto Internacional de Maputo, the airport of the capital of Mozambique, to the city center. I looked out of the taxi window, and even though I was thousands of kilometers away on another continent, what I saw had so many identical features to what I had seen in, say, Brazil, or Peru, or Venezuela. The outskirts of many places in South America and Maputo looked almost identical, as if just simply transposed. Everywhere the same colorful painted huts, the Coca-Cola advertising, the small stores full of endless cheap plastic household goods, the street vendors pushing their carts down the road, loudly shouting about the contents of their offerings—from gas bottles to fruit to homemade pastries. I would later see almost indistinguishable scenes in Amman and Islamabad.

Despite an ever-growing collection of travel experiences and journeys, at the beginning of each new trip I would always find myself in awe of how each place awakened my senses. I felt the distinct richness of a country that can only be found in its colors, smells, tastes, sounds, and encounters.

Yet, at the same time, with each new country I would arrive in, an awareness began to open of seeing beyond any differences from the last country, and the country before that. There was just an overriding realization of how the globalization of these places was affecting the way we live and think. A globalized way of thinking, and a globalized world, does not only mean greater connection and expansion but also often results in the loss of any originality and what is innately individual.

I observed across vast distances, from the cities of the Himalayas in northern Pakistan to the dense rainforest of Brazil, to the coast of Mozambique and the deserts of Jordan, that any supposed differences in how people were existing in the world felt, in many ways, superficial. Beyond the customary cultural diversity of food, clothing, religious identity, customs, and language, in many urban societies, people's experiences and perceptions of the world are becoming more assimilated.

The aim of most young people, and often older people too, is almost the same everywhere, regardless of religion or country of origin. It is fundamentally a matter of belonging, of social recognition and social advancement, financial security, and a good job or career. Throughout many countries it is not only life aspirations that converge but also many approaches to problems, and thus the ways of people's thinking and perspectives.

On the one hand, this reduces the pool of potential solutions and results, while on the other, our mental limitations become far more common. The growing uniformity of people's thoughts can be found in the cafés of Islamabad, the offices of Amman, and the banks of Lima. These experiences provide me with more than enough proof of the necessity of visiting the Kogi, and the likely far reach of their knowledge and experiences, as a people who live far from this ever-increasing globalized way of thinking.

The way in which we perceive both our micro- and macro-world has traditionally been heavily connected to where we live and has been formed over thousands of years into what we call culture. Our way of thinking, the self-conceptualizing structure of our thoughts, as well as the thoughts themselves are an essential part of it. We must remember that one of the central revelations of anthropology states that our environment does not exist in an absolute sense but is a model of reality. Different models exist with different results and consequences, and yet even those that are deemed good models have their own limitations and problems, just as those seen as lesser models offer potentials and momentum.

Wade Davis, in his TED Talk "Dreams from Endangered Cultures," speaks of an ethnosphere that covers the earth, that is, a network of the myriad cultures that exist, a web that connects all spiritual and cultural life and is as important for the well-being of the planet as the biosphere, the biological web of life that also covers the earth.[6] This cultural web of life can be defined as the sum of all thoughts, dreams, myths, ideas, inspirations, and instincts that have been discovered and cultivated by the human imagination since the dawn of consciousness.

Just as the biosphere is reduced by the drastic extinction of species, the lasting legacy of the ethnosphere is eroded by the progress of homogenization: the only difference is that this erosion of culture happens much faster than in the biosphere. One of the crucial indicators of this is the disappearance of languages. In relation to the total number, languages are dying out at a much faster rate than animal and plant species. When a language disappears, a perspective on the world disappears, because some aspects can only be accurately expressed, or perhaps even thought, with certain words of a specific language.

We are quickly becoming a world devoid of individual identity and cultural character, one whose imagination and creativity is being diluted and dulled at an alarming rate. What we truly lose in a world of global conformity and sameness will sadly only become strikingly clear after it is too late. What is already clear to me is that as long as homogenization continues, the more a personal limitation or blockage becomes a commonplace one, and the more we continue to create a world that cannot develop and flourish with bold, individually driven inspiration and ingenuity. The modern world suffers greatly from human interference, psychological turmoil, ecological destruction, and the increasing global crisis of societies that lack meaning in their lives, which is not only harming individuals, but is also destabilizing entire countries. Even though we have made great technological advancements, we have little more than superficial solutions to our ever-increasing issues, and this is exactly where the core strength of the Kogi lies. They have vast knowledge of how to care for, and increase, the essence of life that is within everything and maintain its foundations and principles in an incredibly intelligent and acute way. The practice of these astute principles enables not only a different way of being in the world, and interacting with it, but also assists in creating greater technological solutions and forms of structure and organization that complement the living world, and that do not destroy it. This is beyond doubt the Kogi's immense value to all of humanity, as they are bearers and guardians of this wisdom.

It is not just about feeling alive, it is about stepping into dialogue with the aliveness of everything, especially the earth. The Kogi often told me, bluntly, that by observing our way of life, they perceive us as practically dead.

When was the last time you can honestly say you felt fully alive, fully in tune and in balance with the world around you?

That you spent time in pure awareness of the natural world, without activating your analytical mind, simply observing the details and inspiring qualities of nature?

That you allowed yourself to be submerged by the miraculous sensations of being alive?

When was the last time you considered life from another, new, perspective, from a person or being other than yourself, and what might you witness, what might you hear?

When was the last time you truly listened to the silence of life?

CHAPTER TWO

What Thoughts Have You Been Thinking?

Back in Germany I had heard the Kogi Mamo share his message several times over several days of his six-day tour of Germany. Each time left me feeling a different emotion. This was a man not merely sharing potential problems and solutions, but he was also sharing his own experience of living in a deeply profound relation with the natural world and what the suffering of our planet is doing to the ecosystem that sustains him, his people, and all of humanity.

In his talk below, Mama José Gabriel refers to his hat. The Mamo hats are made from tightly knit white cotton, in a conical shape with a pointy top. They are their antennae to the higher realms and are emblematic of being a Mamo. Designed to indicate the lineage and *esuamas* of origin of the Mamo—the sacred places in the Sierra Nevada—the hats have different colored horizontal lines depicting the different climate levels that can be found on the mountains of the Sierra Nevada de Santa Marta. A Mamo is able to read another Mamo's hat for information. Mama José Gabriel speaks:

We are the Kogi, we are the firstborn, and we still think the original thoughts. Today I'm talking about things that we experience in the mountains, in the rivers, on the heights, and in the forests of the Sierra. However, this is not just something that happens in the Sierra Nevada de Santa Marta, but on the whole planet, even in the entire universe. Jaba Seinekun, Mother Earth, is getting worse.

Sezhankua arranged everything, without exception. Everything is in order. The lines on my hat show exactly where snow can fall, where we live, how everything is coherent.[1]

In Colombia we live at the highest point, the Younger Brother lives below. The water flows from the top downward, then it comes back up from the bottom. The people below pollute the water heavily, even though it is the blood of the earth.[2] That's why there is so much drought. The mother of the water, Jaba Nikuitzi, is sick. The water comes from the lakes up in the mountains. My hat represents it all. Everything looks small on it, this is how Sezhankua sees the things, for us though they are very big. When the elders say that something is going to happen, we listen to them.

Diseases arise because something is out of balance. People get sick because they are out of balance with themselves and with the world. The earth is in a great imbalance because we mistreat the sacred places. We are all children of the sacred places of the earth. We are all children of the water, the wind, the trees, the animals, the sun, and the mountains. We couldn't live without them.

What would we drink if there was no water?

What would we eat if the sun didn't make the plants grow?

How could the plants grow if the wind didn't distribute the seeds?

Where would the rivers come from if they didn't flow from the snow-capped peaks?

Everything is connected and everything is alive.

The Great Mother said that we must not extract anything from her stomach or intestines, but we humans don't stick to that, we unearth everything we find. If a person undergoes a surgery, they will never be exactly the same as before, there will always be a difference. The differences are very small at first, but they add up, and strange things come to us.

What is it that makes people act so destructively? It is the artificial thoughts like greed or envy. These artificial thoughts have gotten out of hand and we feed them every day. You say that everyone thinks like this but that's not true. In the past, these thoughts and feelings did not exist. We always knew that there was enough for everyone and that everyone was cared for. You forgot that, you think you always need more and more.

The increasing imbalance in nature shows that the imbalance

of our thoughts is getting bigger. Unknown diseases are coming. Nature does not care whether we have a doctorate and say that we are scientists and know these things very well. It doesn't matter to her. Instead of analyzing her, we need to speak to her, because you are only investigating the visible! Life works so well, but that is because it follows invisible laws. All natives, not just us Kogi, ask for permission before we build a house. But have any of you asked for permission to build the railroad tracks and roads?

People think that it's just about technological development. They think the future consists of even more machines. They think that one only needs to fertilize a field properly so that everything grows well, but they say that because they don't think with the heart. If they did, they would know that with it they are harming the earth. But everyone, even some of us native people who come into contact with a lot of artificial things today, is getting weaker and weaker. The solutions are the original principles, the remembrance of the knowledge of the First Thought and to respect and practice it.

When the Kogi look at the world, they first see the destruction of nature that, for them, is painful to witness and that without question they would like to stop. They are horrified and deeply concerned about it, but their perspective reaches beyond grief and dismay to a diagnosis: the root of all of this is our underlying thoughts.

The Kogi distinguish between "First" (original) thinking, known also as the "One Thought," and "artificial thoughts."

When speaking Spanish, the Kogi often translate the word *zhigoneshi* as *un solo pensamento*. In English, as well as the "One Thought," it also means the "One Thinking," the "First Thought," "yielding," "we work together," "we are related," or "you help me, I will help you." The translation we use in this book is the "One Thought." In order to attempt to approach and capture the richness and depth of what is being expressed by that, we have to take into account that the Kogi do not understand the words "thought" and "thinking" to merely mean something that is cognitive and theoretical, but that also "thought" and "thinking" are feelings and sensations. Feelings and sensations are only derived from that which is alive, and so it goes without saying that thinking, as the Kogi understand it, is based on that which is living, which can be described as life force.

A machine that handles only abstract processes does not think in the same way because it cannot sense, not even when it is equipped with artificial intelligence. This is an important point because, as we shall see, the Kogi distinguish between thoughts that are alive and thoughts that are dead, that is, thoughts that are derived from the living sensation of life force and thoughts that are not.

While they are still in touch with the One Thought (the original thought), we, the Younger Brothers and Sisters, have got caught up in artificial thoughts. The environmental degradation is a clear symptom of this.

However, it is important to the Kogi that this is recognized as a symptom, and not as an independent problem or an inevitable side effect of economic or technological progress, and that it is the result of our worldview. This means that everyone is included, even those who are not primarily involved in the destruction on a large scale.

Arregoces Coronado-Zarabata shares:

The balance between nature and humans is disturbed. We ourselves are the ones who harm nature. We natives, especially here from the Sierra, talk so much again and again about this topic so that people around the world can understand. Nowadays, there are such huge quantities of advanced technologies that destroy nature through their production or uses. There are also technologies that do not rob or harm nature, but that are gentle and kind to it. There is an existing principle to which nature and humans were created. If we create technologies they must follow the same principle, only then will they be alive and aligned to the principle of life force, otherwise they can be very harmful. The same applies to our organizations. If the government believes that a future can be created by extracting and exporting natural resources, if it believes that everyone will be better off, then they are very mistaken. Because it damages nature and we are all part of nature. Natural resources are the lungs of the Mother, she breathes through them. These organs are very important.

Nowadays, one of the biggest problems is that we use technologies that harm nature. I heard that there are ships that operate under water. No whale or dolphin will start living with us on land. Why not? They still keep the laws that were initially laid down by the Mother.

All pollution, excavation of mines, and grave robberies lead to drought. At the moment we think that this is only going to happen here in the Sierra, but it will spread to the whole world. Natural disasters are the Mother's way of clearing up imbalances. Changing thoughts, such as to the One Thought, can reverse this destruction. Together we will help the Mother, we will collaborate with her in harmony, then the Mother will also help us again. The Mamos are very concerned about all the destruction. The change in climate is a sign of the spiritual undernourishment of the Mother. If the earth is healthy, we humans are healthy too.

For the Kogi, "artificial thoughts" are thoughts, ideas, and actions that are not connected to the flow of life, the aliveness of all things. Those artificial thoughts isolate different aspects of life from each other, both on an individual and societal basis, that essentially cause fragmentation within our lives and cause a deep confusion. Without beings, animals, rivers, mountains, and plants living in cohesion, there is no foundation for harmonious action.

The Kogi see the existence of conflicting thoughts and indecision as a clear symptom of our unstable perspectives, whereas they believe there are questions that in themselves offer clarity and balance, and that do not lead to any indecision or doubt. The Kogi give enormous importance to clarifying any potential complications that may arise in a situation before they agree on an action—a process they call "inner ordering."

Our society perpetuates thought structures that ignore life force and its flow. This applies to every political ideology as well as to everyday life situations in which we act automatically based on ingrained or systemic thoughts. We often perceive these thoughts as our own opinions when in fact they are an expression of a collective view, or one that has been handed down to us. Thoughts that are not in alignment with the principle of life will sooner or later become conflicted, and this will inevitably lead to destructive situations.

Artificial, isolated systems are not connected to the universal energy of life and are unable to generate their own life force energy and therefore rely on that energy being supplied from an exterior source to maintain themselves. This is visible in our entire modern civilization, where our electricity is generated by means of largely finite fuels. Without electricity, everything would collapse. From the Kogi perspective, artificial thoughts must be put back in order, because they affect our world to a point of severe imbalance.

THINKING TOGETHER

"*Zhigoneshi* is a word of ours," says Mama José Gabriel. "*Zhigoneshi* means a single thought. That is how we think—in One Thought."

I wanted to understand with greater clarity what exactly is the One Thought.

Arregoces Coronado-Zarabata speaks:

In the beginning we were given the task that we must look after this earth only by thinking one single thought. We call it *zhigone-shi*. There are so many different people in the world, some are natives, some are not. The thought that everything must be cared for was entrusted to us, and it is one all-encompassing fundamental thought.

The principle of how everything coexists is one all-encompassing fundamental principle.

Our home is the Sierra Nevada de Santa Marta; we are four peoples living here and guarding her. We need all four peoples to work together with one single thought, because we complement each other. We differ in language and clothing, but not in thoughts.

We protect everything, from the snow mountains to the beaches. The Sé Shizha (or Linea Negra) encompasses our territory.[3] It is like a spider web for us. That's why the spiders spin their webs, to show that our planet has a structure.

The Linea Negra is the boundary around the foot of the mountains that indicated the Kogi's original territory.

When the Kogi speak of protecting a place, they mean it in the literal sense we would understand, but it also goes deeper than that, as also guarding and stewarding it. Just as the Kogi would protect a place from destruction or pollution, they will also protect a place through ritual, song, dance, and communication with correlating sacred sites.

Arregoces Coronado-Zarabata continues:

This is how the Mother created the world in the beginning: first the lines of the Linea were created from thoughts and energy, then everything else was formed around it. The Linea Negra is very, very important to us. It runs where the sacred places are, where the streams are, where the mountain peaks are, and where the lakes are. All of this is part of the Linea Negra. We only received the Sierra in order to protect it. We have to guard everything that is on earth. All the objects that exist on the earth were put there in the beginning by Jate (meaning father) Teiku,[4] he placed them in a very specific form and with a very specific function on the earth.

All our dances and chants are only there to unite us with Mother Nature. This is how we live together with nature. We are all the same, the animals, the plants, the rivers—we are all nature. We all come from nature. If we think that all the minerals, metals, and natural resources belong to us, then we are tremendously wrong. In truth, we belong to them.

The Kogi view the One Thought as an innate intelligence within nature with which we can either align our thoughts and actions, or not. It is important to note that the notion of the existence of the One Thought only arises in humans, because in nature it is inherently in action and therefore simply *is*. Every natural thing instinctively knows how to be its natural self.

As humans we can either attain and animate this order or block and unsettle it. Therefore, the One Thought is based on a living intelligence and brilliance that at times manifests itself as wisdom in a human being.

Even though the Kogi's manner of expression is strongly related to their habitat, their statements are nevertheless based on clear principles. The word *principle* is derived from the Latin word *principium* and means inception, beginning, or origin. A principle is the origin of something and represents a law that is perpetually valid and thus forms the basis of further laws.

Therefore, when the Kogi talk about the One Thought as a principle, for them it is not a culture-specific concept but a universal law that applies to all people and to all life. The One Thought is a shared sensation of aliveness, of a universal connection that is evolved from what animates and enlivens all existence and the very fabric of life. Everything in nature is alive, not only because it is interconnected, but because it is inter*is* (from interbeing): If there were only bees, they would not know where to get the nectar from. If there were only flowers, the bees would be missing, and if there were no bees or wind, the ecosystem would falter after just one *summer*.[5]

The same applies to humans: we may be autonomous beings, but we are dependent on the structure of life that surrounds us. This does not mean that we should all think uniformly, but rather it directs us to the contrary. Our thoughts should, and must, differentiate, while relating to the same intangible and invisible essence of life: the way a plant and an animal are very different beings in nature, and in their thinking and behavior, but they equally possess an aliveness embedded in life force.

Mama José Gabriel explains:

The earth is not only sick because of cars, air pollution, or plastic pollution, but also when we abandon our principles. We will all come back to *zhigoneshi*, also the Younger Brother will. When he reads this book, he will want to contribute. He will understand how to act and even better how not to, because the principles apply to all of life. We move forward with one single thought. The essence of life is one single thought. The Elder Brother and the Younger Brother both contribute. If we have a lot of different thoughts then everyone does what they want, and then we continue to damage the earth, then we continue to damage the water, and Mukuakukui, the sun, and Saka, the moon. This is not good! We are still doing our work and we will all inwardly align, which is to order ourselves well. If we think together we will all think well, and then the thoughts will still be functioning well.

The One Thought is not an abstract collection of many individual thoughts pieced together to try to bring forward an idea or action, or from simply following commonly held ways of thinking to bring about alignment within society—but it is about becoming aware of and building a relationship to the alignment within life that naturally exists and connects everything. Within this alignment, everything follows a vital and hierarchical order.

HOW DO WE KNOW WE ARE LIVING ONE THOUGHT?

The majority of us have experienced moments of flow, where we feel and sense that everything within a situation is intertwining perfectly, without significant friction or confrontation.

And oppositely, there are times we feel "out of tune" and disconnected with the world, and all our endeavors seem tedious. Often we are met with continuous dead ends where nothing quite pieces together. This flow, that is embedded in our existence, applies to all aspects of our life, whether it's in the way we build our houses, or establish our societies, or form relationships with all living beings.

According to the Kogi, I am not separate from the One Thought, yet I separate myself from it through an illusory sense of division that does not exist. And this is also true of connection. One can never be disconnected; we can only create, or adhere to, a fabricated sense of disconnection. The essential component of the One Thought is its interconnectivity, rather than a demand for a uniform way of thinking.

To adhere to such a prescriptive way of life and thinking,

as has so often happened in history, is to deny what truly connects us to ourselves and others. We blindly uphold commonly accepted ideas, and, even worse, begin to believe those ideas are our own. The Kogi are not asking us to adhere to their way of thinking but to tune ourselves into the One Thought that animates all life.

Unlike the Kogi, we see different areas of life as separate, and therefore try to develop diverse approaches for each area in the form of tools, techniques, or methods, be it health, management, family, sport, nutrition, or suchlike. But the Kogi see One Thought as applying to all areas of life.

This deeper, fundamental perspective also unites many native peoples. Despite their different life contexts and mythologies, what they also often have in common is that they revere the sacredness of earth as a mother. When we think in the One Thought, we are intrinsically connected with health, joy, and fulfillment. At the fundamental essence of nature there is no disparity between my well-being and the well-being of another. On the contrary, each well-being is one and the same and mutually enriching through diversity.

Mama José Gabriel speaks:

When we humans are born, our birth is very closely connected to the earth. If we are in our mother's womb, then we are in the Sé. Sé is the darkness. We emerge from the water, and only then everything else comes into being, and it was the same with the earth itself. She was in darkness too, at the beginning, before water and before everything else came into existence. Life, every life, emerges from darkness and water.

If at night we dream that something will happen to us on a journey, we consult the darkness through the Oracle before we leave. It is the thoughts themselves that guide us. We can never be separated from our thoughts. Today I told you what we're going to do tomorrow. The thoughts already know what will happen tomorrow because the energy always already exists. Before there were humans in physical form, they existed in thought. Thoughts don't die, they live on.

Most of us have forgotten the importance of thoughts in our everyday life; we are no longer aware of their immense significance, apart from, perhaps, a superficial search for the right "mindset" that promises success. Unlike a mindset, which is an operating system located solely in our psyche (of course also influenced by our environment), the world of thoughts for the Kogi exists both inside and outside of ourselves.

For example, when the Kogi bought back a piece of land from the Colombian government that had not been inhabited by them for over five hundred years, the first thing they did was to cleanse it from foreign thoughts that linger on the land, even though the old owners have left, to reconnect the land with the thoughts of their ancestors and thus with the thoughts of Jaba Seinkun, the Mother Earth.

For Kogi, Aluna, the world of thoughts, is more real than

the world that we perceive with our eyes. The famous ethnologist Gerardo Reichel-Dolmatoff has attempted to describe it as follows: In Spanish the Kogi translate the word *Aluna* to *espíritu, memoria, pensamento, vida, voluntad, alma, intención*; in English: spirit, remembrance, thought, life, will, soul, intention. The word most often used is *pensamento*, "thought" in English, which also relates to the words *energy, idea, feeling, emotion,* and *sensation*.[6]

Aluna also means "original" and "invisible." The act of expressing one's thoughts to a Mamo, which they refer to as "confession," is called *alúnayiwási*.[7]

Aluna is best described by using examples. The Kogi often make a *pagamento*, which is an offering that can be made both physically and in the form of thoughts. It is always an offering to the earth, or to a certain aspect of it to balance what we humans have taken from her.

If the person is not at the corresponding location that requires balancing, a pagamento can be made in Aluna. If a man wants to build a house and envisions it in his mind, he perceives it in Aluna. He also sees the Aluna of the house, the idea. Aluna therefore defines thoughts in their entirety—the thinking as a process, as well as the thought in the sense of a complete idea. The process of creating through thoughts in Aluna is a central theme in the myths of the Kogi.

All visible forms have their invisible equivalents in Aluna. Just as Sé is the world of pure existence, it is also the ordering and form-giving principle that represents the place where all feasible possibilities are stored. Even though Sezhankua and Seinekun created the earth, it is the responsibility of us humans to maintain and nurture it by cultivating clear and pure thoughts.

One day, I witnessed the Kogi discussing whether or not a village they had just recently returned to should be open to tourists. When the Mamo who was responsible for that village heard of this idea, he held discussions and consulted the Oracle in Aluna, and it was decided from there that this misinformed thought must be given back to the fathers and the mothers of this confused idea, and just like that, the question of welcoming any tourism had quickly been dismissed.

CHAPTER THREE

Do You Know
the Stories Well?

The door of the plane opened and the hot humid Caribbean air hit me like a brick wall. It was shortly before ten at night, and the temperature was still over 30°C. We had landed in Santa Marta, at the foot of the Sierra Nevada mountain range, a place, our taxi driver had told us as we navigated our way from the airport, that is the oldest city in the Latin American continent.

After finding a hostel to stay in, we strolled through the dimly lit streets. The wind blew strong and hot in our faces, and gusts of air kicked up dust that got into every pore, the dryness around felt suffocating. We later found out it hadn't rained a single drop for almost three months. Not many people were out at this time of night; the streets felt eerily melancholic and foreign. I had a sense that the stories that unfold here are a world away from what I knew back home in Germany.

We were less than one hundred kilometers from a place that had held deep intrigue for me for a long time, the Ciudad Perdida, the Lost City, a place that many people believe to be the mythical city of El Dorado. I later found out from the Kogi that the actual name of the city was Teyuna, and that it was far from being a lost city at all. It is alive, and sadly it was being ruined, fast becoming a place of tourism, and the only thing lost is respect for its current significance to the Indigenous people of the Sierra.

Santa Marta is only considered the oldest city in Latin America within our modern understanding. The Palestinian author Mourid Barghouti writes that the easiest way to disempower a people is to tell their story and start with the word "Second." This creates a narrative that does not start at the beginning and essentially places the references out of context and falsifies them.

The Colombians begin their own history with the arrival of the Spanish, and therefore deprive the Kogi of ownership of their ancestral land. The city of Teyuna is of no spiritual significance in the world of the everyday Colombian; I would guess that almost none were even aware of its true name. Although they might have some awareness of the existence of the cultural world the Ciudad Perdida belongs to, they have no affinity to it. Even though the territory of the Kogi starts just two hours from Santa Marta, most people who live in the city have never been or even considered going—for them it is as far away as Germany, or even farther. Jate Teyuna, the father of the wind in Kogi cosmology, is the guardian of the "lost city" of Teyuna, but just as the city is deemed lost by the Samarios, the name given to the inhabitants of Santa Marta, so too have they forgotten the Guardian of the Wind. For the Kogi, however, Jate Teyuna is alive, and he is still very much present and has been since the beginning of time.

The taxi driver from the airport also told us that in Santa Marta it is forbidden for two men to ride together on a motorcycle. The reason for this—too many people have been shot by passing armed men on motorcycles. We are told that the situation has improved, but in the same breath it is mentioned that all levels of the city government are still run by those put in power by the paramilitary, who in turn serve the two or three most powerful families of Santa Marta. Crime is high and people disappearing is common. It is reported that over eighty people who had campaigned for human rights, environmental protection, Indigenous rights, and freedom in general disappeared in Colombia during a two-month period alone. When someone disappears in Colombia, it almost always means that they are tortured, murdered, and buried somewhere. These cases usually remain unsolved and the bodies are never found or, on a rare occasion, found in the middle of the next rainy season, when the floods erode the earth and bring to light all kinds of ugly crimes.

How the Samarios view, and what they think of, the Kogi has not changed much since the arrival of the Spanish. Many of them still see the Kogi as primitive, or at the very least as peasants without rights. Yet, they live just two hours from this city of apparent danger and conflict, with a culture that is alive, with cities that are not lost, and, in their own language, without a word for enemy.

For most Colombians, the rich culture of the Tayrona, who were the ancestors of the Kogi and lived on the same land prior to the arrival of the Spanish, is distinctly different from their own everyday life. As mentioned, there is a strong suspicion that the Ciudad Perdida was the legendary El Dorado. Yet, the only real connection that upholds this theory were the Colombian grave robbers who, because of how many there were, laughably even founded their own trade union. Without any respect and with little aptitude, they unearthed golden figures made with great artistic skill from old Tayrona graves. They would sell these on the black market to middlemen who would then offer them to interested art collectors from all over the world. Aside from this tale, the Tayrona are mentioned rarely, only by a handful of tourist agencies or adventure trip organizers.

Within the Parque Nacional Tayrona, posted signs reference the life of the Tayrona, and their rich culture, purely in the past tense: "Many centuries before this beach became a tourist destination, the communities of the Kogi and Arhua-cos, a brother people, came here to make *pagamento* to the Mother Earth in order to maintain the planetary balance."

I am left with the impression I am reading about a museum artifact, something that is extinct and resigned to the ancient past, while in fact the Kogi still make pagamento, and only do so in the mountains and not the coastal beaches because they are not granted access to the coastal sacred places.

THE KOGI SAY THAT A NATION WITHOUT MEMORY OF THE origin is a dead nation. They share with us what happens when complex systems that inform our everyday actions lose their meaning and significance and become dead actions, to the point where they become automatic and unquestioned systems. They may still entail specific goals, but these goals are not embedded in any real meaning. A lack of meaning or memory is an exceptionally modern dilemma that differentiates us from the Kogi. We conceive meaning as something that has to be "found"; the absurdity of this idea should make us pause for a moment. If we are yet to find purpose, has our life been meaningless up until now? And what does it say about the actions we take in our everyday life? What if meaning is not something to be found, but is something that is cultivated and brought to life?

The only way we can truly understand how much we see the world as relative to ourselves and limited to just that, is by stepping out of a situation and reflecting on it. This is the only way to discern what is reality or illusion. This is true in many aspects of our lives, from business to family, relationships, health, and how we can best grow internally.

Today, it is important to recognize that Indigenous people who still live according to their original traditions are the few remaining examples of those who live outside of a globalized way of thinking.

In contrast to many other Indigenous peoples, the Kogi have succeeded in preserving their original traditions to a high degree and have maintained them, mostly away from the influence of modern Colombian society. Only recently have items such as rubber boots, machetes, solar flashlights, and smartphones slowly found their way into some parts of the mountain range while other areas have remained virtually untouched by modern products.

The Kogi way of living is not governed by money or industry, or markets or states, and therefore does not have the accompanying mental obstructions or limitations. At the same time, they have preserved expanses of wisdom in their thinking, which we could assume we also knew at one point in our ancient history but have since long forgotten. But first I want to share with you how we even arrived with the Kogi in the first place.

"HEY, HEY—VEHICLE IS NOT BROKEN, GOOD AIR CONDITION-ing, cold, cold, really very cold!" screamed a ticket agent for a local bus company. Soon a rival agent piped up, "Best driver, most comfortable, very cold!" and on it went, a chorus of intensity.

Slightly bemused and bewildered, we were standing at the bus station in Santa Marta a few days after our arrival, trying to discover which of the different bus companies would be best for our four-hour trip to Valledupar.

We picked a bus company at random, paid the ticket fee, and waited; unfortunately, the minibus we chose ended up leaving one hour after its specified departure time, the air-conditioning cooled the vehicle down to what felt like a rather frosty 12°C, and from the loudspeakers the Vallenato, the typical folk music of Colombia and especially of this region, was deafening. We should have expected this musical roar; after all, we were on our way to the birthplace of Vallenato, Valledupar.

The next morning, having arrived in Valledupar, we neared Casa Indígena, the office of the Organización Gonawindúa Tayrona.[1] Before our departure from Germany we had arranged to meet Mama José Gabriel there that day. It was Sunday morning and the streets were deserted. We only saw a few cars driving around and a handful of pedestrians. Carrying cumbersome backpacks, we pushed open the heavy iron gate to the courtyard of the Casa Indígena. There was nobody there, just silence. We noticed a big mango tree in the courtyard and decided to take shade under it and wait. It was slowly approaching noon and would likely be 40°C before

long. Under the sweltering hot sun and humid air I was trying to keep my mind clear, considering our options of what to
do next. I found my tired mind transfixed by some dry leaves
blowing through the courtyard and the way they interact
with the wind, dancing through the air. I started to consider,
what will our stay in the Sierra with the Kogi be like?

Abruptly, I was brought back to our present conundrum
by a group of Arhuaco entering the Casa, but they quickly
disappeared into the back building of the property. We
waited. After about an hour, a Mamo of the Wiwa came out
of one of the rooms to the side of us and we asked for Mama
José Gabriel.

"Yes," says the Wiwa, "I met José Gabriel this morning at
a meeting; now he is already on his way back to his village."

We were confused and our thoughts ran fast: We had
missed him. Were we too late? Had we been wrong about the
time? Had he forgotten about our arrangement? We did not
know what to do. We didn't know the way to the village and
had no other possibility of getting in contact with Mama José
Gabriel, because most of the Kogi have neither a mobile phone
or telephone. The Mamo of the Wiwa looked at me very intensely, and when he noticed the beginnings of my emotional
crisis, he asked why we wanted to see Mama José Gabriel and,
somewhat harshly, what our intention was with the Kogi in
the first place. After explaining our situation to him, he asked
me to wait a moment.

An hour later, an old, rusty car pulled up outside the Casa.

"I have a friend who will take you to Altamira.[2] There you have to ask for Juan, he can help you," said the Wiwa Mamo in Spanish. We hesitated. Colombia is not a country where you just blindly get into someone's car, least of all if it is not even a taxi.

After a brief moment of reflection we saw that this was our only chance to potentially get to Mama José Gabriel that day. So we decided to quickly buy some groceries at the *centro comercial* across the street, as we were told by a friend that the Kogi expected us to bring as much food as we would consume during our stay with them—by doing so we would not create an imbalance. We bought as much as we could possibly carry, about twelve kilos each of rice, potatoes, canned fish, lentils, and oil. We began to thank ourselves for keeping the rest of our luggage light and stayed hopeful that there would be mules up in the mountains to carry our heavy loads.

Dust showered us through the broken car windows, and dreamy Vallenato music accompanied us along the bumpy dirt roads. Diomedes Díaz, a well-known Vallenato singer/songwriter, purred "Tiene razón ella, tiene razón en ciertas cosas, porque deberás yo reconozco de que si soy mujeriego."[3]

The dirt road was lined with miserable looking cows huddled together in the only shade they could find, under the few spartan trees, hiding from the blazing sun. The grass pastures were completely withered and yellow at this time of year, at the end of the dry season. Our destination, Altamira, is the center of what is left of the lands of the Kankuamo and their culture. The whitewashed stone houses and small church at the center of the village could, however, also grace any other Colombian provincial town. At first glance, there was little to suggest that there were any Indigenous inhabitants living there; everywhere I looked sat men wearing cowboy hats.

We were passing through the narrow, winding streets of Altamira when suddenly the driver stopped in front of a house, where a large family was sitting under a canopy. Our driver talked to them briefly, and one of the men stood up and greeted us. All of a sudden, all eyes were on us, a couple of confused and exhausted foreigners.

"Is Juan here? We would like to talk to him," I asked.

"No," he replied and looked at us curiously. "He is in town. What do you want?"

We told our story and the expressions on their faces lightened up. They said they would help us.

Half an hour later, two teenagers on motocross bikes were standing in the front garden. They said they would take us to Tarquilla and from there we would go by mule to the Kogi village, Machukúmake. The villages are spread out over a vast distance along the steep mountain range. Each village has between two hundred and fifteen hundred inhabitants. In total there are roughly fifteen thousand Kogi. The villages higher up the mountain are more secluded and more traditional, while the villages lower down are more in contact with the outside world. From many of the villages the Caribbean Sea is visible in the distance.

"Yo llevo la embra,"[4] one of the boys said to the other and asked my partner to get on. She jumped on, as did I on the other motorcycle, and our journey to Tarquilla, the last frontier of Colombian society, had begun.

The road to Tarquilla had more in common with a dry river bed than with an actual road. The journey there was one of the most absurd of my life. Both of us were sitting on different motorcycles behind the driver, each with a big, now full, backpack on our back and a small one in front of our chest, while clinging on for dear life as we swerved around bends and bounced through potholes. At points we were traversing sandy ditches up to a meter deep and bumping along fields of rubble. The two drivers turned up the throttle with force to make it up the steep mountain, and a good half an hour later, after many near misses and our stomachs fully churned, we found ourselves on a bit of paved road again. It was a sudden surreal pleasure, and it didn't last long. After only a few kilometers, the road turned into a dirt track once again, and we held on tight, until we eventually reached Tarquilla.

It was late afternoon, and due to the higher altitude the temperature was very pleasant despite the strong afternoon sun. The relief of having made it so far took over, and we were finally able to breathe and take in our surroundings.

We put our luggage down and I decided to search for the Mamo we were told would be waiting for us with a mule to take us farther up to Machukúmake. It wasn't long before my creeping fear became a reality. We were met by a man from the Kankuamo who could not have been younger than seventy years old, and much to our dismay there was no mule in sight. We had no way out—we would have to carry our overloaded and bulky bags ourselves.

The man went first, and we followed. He was a very good walker and moved at a brisk pace. It was not easy to keep up with him; at times we were moving faster than our bodies agreed with. He never once looked back to see where we were or even took a pause. At some point we had left the dry, withered landscape, and as darkness fell, we walked through a small forest with rushing streams and lush greenery. We passed some Kogi men who were out wandering with their mules and cows. We had a clear sense that we had now crossed a threshold between territories.

We had left one world behind and entered another. Actually, we were still in no man's land, in a space between Colombian society and the Kogi, between the parched, overgrazed grassland and the green hills of the Sierra. The shift in landscape and people was visible everywhere and reflected a profound transition, which was also highlighted by the unique, constant, and inquisitive look the Kogi gave us as they walked along the path. After the last rays of the sun had disappeared, the mountains were covered in a deep blue light. Dusk came, and at the same time the world that we knew, and with which we were familiar, faded away behind us.

Just as darkness set in completely, we arrived at a small

village, undoubtedly Machukúmake, and to our deep satis-
faction, Mama José Gabriel, the Mamo who had invited me
there when we were still in Germany, stuck his head out of
the *nuhué*, the Kogi ceremonial house, and looked at us:
"Ah, there you are. I didn't know if you would come today.
You have changed the date before. There's your hut in the
back, you can hang your hammocks there."

And with that, he disappeared again, into the darkness of
the ceremonial house.

OFTEN, WE IN MODERN SOCIETY ARE UNABLE TO COMPRE-
hend the thought processes or actions of Indigenous people
because they live a life so different from our own. This is particularly true in the case of the Kogi, but this works both
ways. The Kogi do not understand why we do things the way
we do. We have two completely different ways of existing in
the world. Take Tarquilla and Machukúmake, for example,
two places that are just an hour's walk apart. Why is it that
the Kogi and the Colombians, who live so close to each other,
perceive the world so differently?

A distinct characteristic of the Kogi, and their neighboring peoples, that I have noticed time and again is that they
often begin longer conversations with the origin of the world
and expand from there. So I was fortunate to hear the story
about the essence of the earth a dozen times.

In the words of Ade Wiwa Mama Ramon Gil Barros:

In the beginning there was only Sé. There was no earth, no time, no air, there was nothing, only Sé, and it seemed that it would be better if there was matter. In Aluna, Sé created Sezhankua and Seinekun so that they would create the physical earth, because Sé itself cannot move or act, it is simply being. So Jaba Sé gave them a thread of thought, first to mark out the boundaries of the territory at the mountain peaks, then the middle layers of the Sierra, and finally a third thread to define the borders of the mountain at its base. This last line was called Linea Negra.

Different plants and animals should live in each of the three zones. The center of the Sierra was called Gonawindúa. *Gon* means to be born or to build, and *gonawin* means the movements of a child in the womb of its mother. *Du* describes life for many thousands or millions of years, and *dua* means the first seed of matter. It was when Sezhankua and Seinekun sat on Gonawindúa and looked at the Sierra Nevada that they recognized many animals and plants as the seeds of life. Then, however, they observed that a place for the sea and one for the earth was required. Back then, the Sierra was still bare, the soil on the earth was missing.

Jaba Sé told Sezhankua to take the black soil as his wife, but Sezhankua found her too dark, and he didn't like her, he wanted a woman that was more light. At first he married the white soil, but she was not fertile. Then followed the brown, the ocher, and the red, but none of them were fertile and none gave birth to offspring, neither did the sandy earth. Thereafter Sezhankua married the earth of black sand, then the green, and then the yellow.

So overall he married eight different women in a row. There was only one left, the black earthy one, for there were nine sisters.

Sezhankua said that he would take the black soil as his wife, but all of his brothers were angry. They said, "He has already married eight sisters and now he wants the ninth. We will not allow that. He'll have problems with us."

So the brothers took the black soil and locked her behind seven doors with seven locks. Sezhankua arrived at every door and played his flute and sang seven different songs. However, he couldn't open the doors.

Then Aluawikuu came and said to him, "Brother, what are you doing here?" Sezhankua replied, "I will free the black soil and marry her. So far I haven't succeeded because I don't have the power to do it." Aluawikuu then asked him why he wanted to marry her at all.

Sezhankua replied, "If I am not with the black soil, there will be no fertility in the Sierra and the world."

Aluawikuu saw that Sezhankua was right, he then said, "Lend me your flute." And he started to play. He played badly out of tune. Sezhankua saw this and thought that he himself could play much better than Aluawikuu. However, after Aluawikuu played the seven songs, the gates opened. He took the black soil's hand and passed her to Sezhankua as his wife.

Since then, she has been our mother, the earth. We call her Jaba Seinekun. That is why we have to take care of her and live up to the responsibilities that our spiritual fathers and mothers have left us with.

Some time later there was trouble. The brothers went to Jaba Sé and said, "Sezhankua took the black earth away. We will pursue him and correct this again."

Jaba Sé said, "Good, reprimand him."

Meanwhile, Sezhankua watered the black earth. This is the reason why yucca, bananas, and other crops grow very well in places where there is black soil. Unfortunately, it is now also the case, that wherever the earth is not left in peace, many problems such as landslides or crop failure occur. That is why the old and wise Mamos say that the conflicts that happen today between the people stem from the earth not being left in peace. They also say that we shouldn't be arguing with our brothers.

SCIENCE AND THE KOGI VIEW

When I tried to explain to the Kogi how we view the earth in general in the modern world, they were unable to accept this way of thinking. I explained our modern worldview to the Kogi: that the earth is the result of the Big Bang, which set in motion processes that supported and formed the many creations of the universe.

I shared the commonly held belief that our earth is essentially a random happening that has no central consciousness guiding it and is the mere consequence of physics—and that the same applies to the creation of life; that we are all simply fortunate that the temperatures on earth were just right for biological processes to emerge in water, and it was essentially random mutation that led to the existence of all species today.

For the Kogi, our view is a factually incorrect way to see the world.

To view the earth, and specifically nature, as a conscious living being that is in a state of constant evolution has, in more modern times, not been easy for wider society to accept. It is a view that has more often been met with resistance and derision. The idea of the earth being inanimate has only been widely believed since the emergence of the Scientific Revolution, a few hundred years. Nevertheless, in the twentieth century, there were some scientific attempts to rigorously examine this viewpoint, and also to counter it. The most famous of these attempts is the Gaia hypothesis presented in the 1960s by microbiologists Lynn Margulis and James Lovelock. This hypothesis states that the surface of the earth and the biosphere on it can be regarded *as* a living being, provided that the totality of all organisms creates and maintains conditions within a dynamic system that enables not only life but also the evolution of more complex organisms. This hypothesis is based on a definition of life that characterizes living beings as those with the ability to self-organize.

However, one word was particularly important for these two scientists: the small word *as*. For Margulis and Lovelock, the earth is only *like* a living being, not actually one. Lovelock emphasized that he was talking about a living planet, but that this should by no means have an animistic connotation. He did not assume that the earth is sentient or that stones are capable of acts of will. He perceived the activities of the earth, such as the regulation of the climate, as merely automatic processes that occur within a certain system and not as a conscious act. Lovelock stated that he respects those who found comfort in the church as much as those who did so in nature and would offer their prayers to Gaia.[5]

Margulis and Lovelock's use of the word *as* created a narrative that seemed new and innovative but failed to accomplish one important aspect: to transform the earth from an object into a subject within our understanding.[6] They therefore essentially conform to the idea of an inanimate, object-like nature.[7] To assume that earth is an object fundamentally prevents three things:

Firstly, it prevents any kind of direct experiential perception and impression of the aliveness of earth, and therefore a direct relationship to it. (It may be that some of us are gifted with perceptive abilities, but in most cases in this scenario, that which is sensed would still not be considered the life force that the Kogi speak of.)

Secondly, it prevents the understanding that a living intelligence governs the processes of nature and that it is not a complex but still mechanistic biological system.

Thirdly, it prevents a scientific approach that is based fundamentally on the life force of the earth. Margulis and Lovelock regarded any attempt to think in this direction as esoteric and thus academically unsound.

Why is the concept of a living, consciously aware earth so

utterly absurd in the view of millions of people in whose lives it has played such a vital role over thousands of years?

There is still no scientifically recognized approach that sees the earth as a subject. Altering this viewpoint among the wider masses would have obvious long-term positive consequences. What if we shifted the question, and instead of asking how much life force exists in the context of the earth, ask how much life force exists in our modern mechanical and technological world?

For the Kogi, the earth is a living being just like a human, its vitality is directly and fully felt and experienced every day, because nature is everywhere: we breathe it, we touch it, we eat and drink it and walk on it. Therefore, it would be misguided to talk of a Great Mother in a mystical sense, as a kind of deity, but more accurate to speak of her as an undeniable experience that is as real as something tangible in nature that you touch. We would never consider asking a baby if it "believes" in its own mother, even though, very much like our relationship with nature, a baby's very survival and development is derived from the reciprocity of this relationship, as if the mother and baby are an extension of each other.

However, we must be clear that this touch, or interaction, is experienced mutually: by us and nature. Within nature there is a consciousness that senses and experiences our touch. Grasping this awareness dissolves the understanding that earth is just an object and allows us to establish an emotional relationship. Only when we love something or someone are we then willing to protect it instinctively. Clearly, it is not enough to just rationally acknowledge the need to protect the earth to really initiate profound change. Only when we experience the earth as a living subject will we be inherently motivated to protect it, just as we would our own mother.

The highly regarded ethnologist Gerardo Reichel-Dolmatoff writes: "I truly believe that the Kogi, and many other traditional societies, can greatly contribute to a better understanding and handling of some of our modern dilemmas, and that we should consider ourselves fortunate to be the contemporaries of a people who, perhaps, can teach us to achieve a measure of 'balance.'"[8]

Of course, there are individuals within modern society who are aware that the earth is alive and weave this knowledge within their own lives as much as possible, but they are often swept up in a sea of subcultures. Their point of view falls into the category of a personal preference or a theoretical exercise, just as some like to visit yoga festivals in their free time or meet up to play a sport. The relevance for "real life" or professional context appears rather limited for most.

In many parts of the Western world, we don't have a culturally intact Indigenous people living in our proximity as a point of reference that would allow us to experience an entirely different way of thinking. But Colombia holds another reality; the inhabitants of Santa Marta, Valledupar, and even Tarquilla are on one side, while the world of the Kogi is on the other. Nobody in Tarquilla speaks more than maybe ten words of

Kággaba, the language of the Kogi. The way of life for the people of Tarquilla is fundamentally different from those who live in Machukúmake, and those farther up in the Sierra. They have copied or adopted almost nothing from other peoples, not even ways to cultivate their fields more sustainably or efficiently.

However, the possibility of dialogue between cultures today, in the twenty-first century, is not only a question of geographical accessibility, but also one of the willingness to engage with those deemed as "other." The people in Tarquilla, for example, live within walking distance of the Kogi yet know almost nothing about them and have no interest in changing that. They are aware of how the Kogi tend to their gardens and take care of their animals and that, in contrast to the gardens and pastures of Tarquilla, there is no overgrazing or extensive drought. Yet they refrain from the obvious: asking the Kogi how they manage to do this. Is this simply due to a general unwillingness to change? Maybe.

Is it because most Colombians believe that the native people are primitive and that their knowledge has no value? Almost certainly.

To reveal an honest point of reference outside of our own boundaries, two things are necessary: the willingness to explore unfamiliar ways of thinking, and also the possibility of accessing direct experiences.

Walking in the territory of the Kogi is an experience of melting into deep time. The glacier-covered mountains feel eternal, as the light of the afternoon sun is reflected on them.

Time moves here, but at an unfamiliar and unrecognizable pace. I feel the presence of present time, yet, spanning a multitude of decades, of centuries, of millennia. There is simply no difference between centuries in the territory of the Kogi. The same water flows down the mountains, the same trees grow, the same birds fly, and, most important for the Kogi, the same stones lay on the ground.

Not much has been changed, and those changes that have been made were in accordance with nature and her principles. Walking into a world that has been guarded and respected for longer than I can comprehend is something that radiates from everywhere I turn; my senses become engaged and alive to the harmonious beauty around me. I feel I can concentrate better, I feel calmer and more focused than in the world outside the Sierra. It is not just the tranquility that many other places in nature emit, there is something else. The Kogi territory feels to me truly alive and awake. The same sensation and awareness duly becomes alive in me, one that grows the longer I am receptive to it.

There is a difference to walking through a territory and walking on a land that has been regarded as a counterpart of humanity, and as kin. A culture and practice of deep mutual nourishment has changed the land, and landscape, without physically altering it. Even though the forest around me is primarily Caribbean rainforest, it seems to have been guided by humans, rather than manipulated by humans, in a way that is not destructive. There is a human effect on my surroundings, just not one that is degenerative, but rather it is regenerative.

As I entered the territory governed by a four-thousand-year-old wisdom tradition, it was the first time I experienced the positive consequences of a regenerative worldview. It takes a lot of time for a mutually beneficial nourishment to fully unfold in a territory. Nature, and the earth, are giving the Kogi everything, and in return the Kogi are giving to the earth. Even land that has only recently been bought back from the Colombian government, it was previously taken from the Kogi due to colonization, feels different and exudes a special quality. However, there is still a distinct difference in what I felt when standing on a piece of land that had been in the guardianship of the Kogi for twenty-five years, as opposed to one that had been under their guardianship for four thousand years.

The land feels devoid of any thoughts of a power struggle or domination. It feels almost impossible that the concept of exploitation exists when surrounded by the pristine waters of the small streams coming down from the Sierra. Even though I am in a new climate, a new country, and a new culture, I find myself sleeping less. It seems that there is more energy in my body than usual, and not necessarily in a sense of more active-ness, but rather in the feeling of being deeply nourished and welcomed by a land that knows that people here are witness to it, they respect it, and they give back to it.

THE STORY OF THE TREES

Every night the Kogi gather in the middle of the village at *la loma*, the hill where both men and women sit together and listen to the elders speak. And every night, they listen as the tales, the myths, and the stories of the beginning of time are told, and they say to me: How can we know the world if we don't know her story? How can we take care of the plants if we don't know where they come from? How will we look after the animals if we don't know who they are? We must always first know their stories . . .

The elder Mamo starts to speak as the stars gradually emerge in the sky, as everyone gathers around the fire:

Let us begin at the origin, in the time when we were all Aluna Kág-gaba [spirit people]. In the beginning everything was one living being. Everything had life and everything was alive. At that time there was a tree called Mashila. The Great Mother used this tree for building houses and for cooking. At that time, however, things did not consist of matter as they do today, they existed only in spirit. The Mother had decided that in the future many trees would be needed in order to build houses, to press sugar cane, and to build canoes. Therefore, our spiritual Mother observed nature very extensively. In the time of Aluna, trees were not felled with axes, but only through spiritual work, because everything only existed in the immaterial world.

Later our own spiritual Mother created nature. She created areas to plant trees and bushes. After she had done this, she sat down and observed what the plants were feeling. Then a Gunama [a Kogi who does not have a spiritual or leadership position] saw a large forest on a hill and went to cut the trees down. There were many big trees, and before he cut them down he began to clear the undergrowth and remove the small branches. When he cut them off, they felt pain, it hurt them a lot and they cried, because they too were alive. But after he had cut down the trees on the wooded mountain, the next day all the trees were standing upright again, just as they had originally been. Our spiritual Mother asked a wise Kwivi [a Mamo apprentice] for advice. He knew the order of all things. The others who were not Kwivi did not want to stop cutting down the trees, but they had to because the trees kept complaining. There were also spirit trees called Jate Kalashé. In order for a person to cut a tree or grow something else in its place, they have to first be granted permission by receiving a stone called Kalguakwitsi. This stone is given to a person by the father spirit.

Mama Jacinto Zarabata adds:

It may be that the Younger Brother suddenly comes here and wants to tell us that in the past there were even trees growing on the snow-covered mountains. Then all of a sudden they may want to start planting trees there again because they think they know better than nature. But it is us who still know the original story. Our Mother left us the snow-capped mountains and they had almost no forest on them. This is exactly how our Mother came to understand it through her studies. Nature also had its Mamos, just like us; we also have a Mamo for every place.

In the beginning of time, when all things that now exist were still only immaterial beings in the spiritual realm of Jaba Aluna, there was one Mamo who was called Mama Shisha. There also existed a Mama Kasouggui, Mama Kazuiku, and Mama Numashi. The latter of the three said, "In the future, I will live very far away from the Sierra Nevada. Although I will be far, I will communicate from there, the place where our father is nourished and given his strength."

We know the ancient story: "In heaven there are no trees, because they, the trees, have decided that they will not stay there. They replied to the Mother that they would be here on earth to serve in this world; in the fire, as a stake in the fence, as well as for the construction of the houses." They said, "When the men decide to come and get us for any purpose, be it as a guardian or to serve as doors or windows, we will continue to serve them in the future, although they cut us down and collect us, even if they use us for the small bench we sit on, or for the big bench that is placed in the nuhué."

So they called all the trees and said, "What do you want to do?"

And one of them answered, "I will serve as wood to make a good sugarcane press from."

Another said, "I will serve as a guardian of the rivers and protect the waters of the streams." That is why today we must not cut down the trees that are on the banks of the rivers and at the sources of the streams. Their father is Mama Kasouggui.

I know that the Younger Brother also makes use of the material of the trees to make seats, tables, and many other things. If they, the trees, had not agreed back then to be used for all these things, the Younger Brothers could not use them today. However, there used to be a lot of forest in the lower part of the Sierra. At the same time, on the snow-covered mountains, such as the *esuamas*, very few trees would be growing for us, the Elder Brothers, to use. And when we spoke with Jaba Aluna, we told her, "If you decide that there shall be forest in some places, then also the places which you define as sacred places must have forest also." Then the Mother affirmed that this was the truth. The Elder Brothers said that the right place for great biodiversity was in the lower parts of the Sierra, below the place that is called Pueblo Viejo [below one thousand meters altitude]. So the Mother asked, "Great son, why do you want there to be forest?"

He replied, "In order to have enough animals, monkeys, wild boars, armadillos, and turkeys, so that they can serve as food." When the Mother realized that this was necessary, she created everything according to their request.

Mama Luis Noevita confirms:

Yes, that's how it was in the beginning, all things were living beings. Trees did not exist in material form, but only in the spiritual realm. Now we see the movement of the trees when the wind blows, but it is their spirit. Now, when the rain falls, we cannot hold it in our hands, it just flows through. The same is true for the clouds. Back at that time a sign appeared; it was called Azkalkwa. There was no dawn yet, it existed only in spirit.

Jaba Aluna observed nature deeply and said, "It is very exhausting for me now to let nature grow, but when I worked here with the Zhatukwa, the Oracle, it seemed that one day there would be no more respect for the nature I have given life to."

Aluna Jarleka then said, "How do you think it is possible that one day we will not respect nature, or not even your son Mama Kasouggui?"

And she spoke, "In the future many things will appear that do not yet exist. Once these things will have materialized, I will have retreated together with my son Kasouggui."

Aluna Jarleka responded, quoting more of the Mamo story, "Things have not manifested into form yet, but when the time of *itakalkwa* [nine worlds] comes, a specific stone of nature will appear."

In the time before Jaba Aluna retreated she said, "You, Aluna Jarleka, know very well what Mama Kasouggui said: the trees will serve everyone. I must give you something like a book, like a dictionary, where the testimony of my words will be found. When you begin to cut down the trees, you will always remember my words. Before you start to cut the trees, you will mentally do some kind of anesthesia for the Father of Nature, Mama Kasouggui. This way the trees will not feel any pain. Because they, the trees, will also manifest into material form in the future, just like a human being. In the future they will have the appearance of a trunk; it may seem that they are lifeless, but they will have life, they will have water, and every year they will give fruit. They will do this, just as a woman gives birth to a child every year. So when you are looking for firewood, make pagamento and communicate with Kasoug-gui. If someone wants to build something, they must consult Jate Kalashé. If they remember to do these things, only then will nature never end."

Jaba Aluna, our Mother, gave us the story of how we take care of nature so that it will always exist exactly how she entrusted it to us. She left a law for each group, for every people. When we plant things, when we cut down trees on the mountain, we make pagamento for Kasouggui. In this way we still follow the law that was left to us by Jaba Aluna. It is true what Jaba Aluna says, that the forest or the water or the river gets sick if we forget to make pagamento. When a heavy rain comes, the mountain slides down and many animals that live near the river are killed. These things are already happening. However, in the past there were no large clearings on the mountain, because our Mother had assigned these places to the animals: to the deer, to the wild boars, and to all others as well. Before the Spanish arrived, we all lived in complete harmony, we all fulfilled our spiritual work. At that time, we did not cut down many trees, and above all, no graves were robbed, no rivers were damaged, and there was no chainsaw.

We native people have always made pagamento before cutting a branch, whatever we take in a material way we give back in spiritual way. At that time, no trees dried up and they did not get sick. Down in Mingueo there were three rivers, many trees, and sacred stones. Jaba Aluna herself placed these stones there. It is where we communicate with Kasouggui and Jate Kalashé and Jaba Kalawia before we cut down a tree. There we make pagamento so that the forest does not get decimated, but since the arrival of the Spanish, the Younger Brothers have started to destroy our ancient pottery in the graves and the sacred stones. They use their big machines to build roads at the places where roads should not be built.

Because of this the old Mamos remembered the story of Jaba Aluna: "It is true that one day things will appear that will cause harm!" Now we see that it is true that the story has its origin in the beginning of time and yet is describing what is happening now. The Younger Brother does not respect nature. When we fell or cut trees, we first communicate with the place where Kasouggui lives.

The Younger Brother, however, is not interested in Kasouggui; they don't even know anything about him; they don't even know his name. When we build something, for example, a path, we talk to Jaba Kalawia first. Before we build something, we all meet to order our thoughts. In the past, only the Mamos were allowed to enter the region of Bunkwangega [the Colombian town of Bongá], not just anybody, but only the Mamos. They came down from Makutama to look for wild animals there. At that time there were enough animals. But the Colonos [Colombian farmers without an Indigenous culture] came and behaved very badly, and they did not care if they chopped everything down or if they harmed the rivers. In six places, for example, there was sacred pottery. They cut down and removed the trees that should not have been cut down, because the trees were also Kasouggui. They cut them down with a chainsaw. They took the wood to sell it somewhere else. That is the same as if you would sell your own father or mother.

By now the moon had appeared in the night sky above us, the fire was still burning, and the Mamos continued to talk. The small village around us was illuminated in an ethereal silver light.

The story of the trees in the Kogi cosmology is central to what we would call a regenerative approach to forestry. It reveals how stories can be alive, and how they influence the way we look at life. It illustrates how everything in nature is alive and interconnected, particularly, in this case, trees. By sharing this story the Kogi do not aim to reprimand us for our destructive behavior, they share it in an attempt to guide us to remember the sacredness and aliveness of all things.

We too can offer a pagamento. A pagamento is the offering of thoughts, emotions, gratitude, songs, dance, or an object to a sacred site and the energies they are connected to. It can be done individually or in groups. It is the practice of being energetically aligned with the earth, as a counterpart and ally, and building a relationship of reciprocity that gives back to the earth for what we have received, such as our food, water, and shelter, and also the wind, fire, water, and darkness.

Even if you live in a city, the possibilities for creating a pagamento are as wide as your awareness. Wherever you witness the earth offering itself to humans or animals is a good place to start, whether the earth is offering a resource or purely its beauty.

Find a place you are drawn to—it can be a tree within the city, a wild plant growing from the cement, a hill in the distance—and bring your attention and awareness to this place, acknowledge this place with gratitude, and open your internal senses. Feel what you are called forth to offer and take action from a space without distraction or limitation. Repetition and continued offering in one space can expand the connection.

WHAT STORIES DO YOU TELL EACH OTHER? AND WHAT STORIES DO YOU TELL YOURSELF?

"¿Por qué ustedes hacen esto? Why do you do this?" I often asked the elder Mamo when he was explaining something to me.

"¡Porque historia dice así! Because this is what our story tells us," he would answer back, and then proceed to enchant me with the accompanying story. I found that the Mamos would always say that we should ask for the stories that exist in our lives, because they determine our view of the world, our thoughts, and thus our future.

"What stories do you tell each other? And what do you tell yourselves?" Mama Bernardo Mascote Zarabata asked me.

"Do you tell stories of life and beautiful thoughts?" added Mama José Gabriel. "Our stories tell us how to treat each other well, how to live well, and how to guard the earth. Only when our original stories are filled with good, do we act appropriately. Our stories tell us where we are going and what our task is, because our stories allow us to still remember.

When we share them, we orient ourselves and put ourselves in order."

The vivid, ancient stories are the backbone of how the Kogi perceive themselves and are the foundations and guidelines of how they take action and make decisions. With astute precision, they apply the fundamental meaning of their stories and the answers and solutions transmitted through them that cultivate and order their relationship to themselves, their community, and the land.

Stories inform us on a psychological, neurological, and physiological level because they generate emotions and feelings and open us to perspectives and impulses for action. Regardless of the reality of historical events, the human psyche will always interpret a story in reference to one's own experiences, that is, in what the historical event means to us. This is an enduring factor of human existence. Mama José Gabriel's question is therefore entirely appropriate: What stories do we tell ourselves about the earth, humanity, ourselves, and our actions? Which stories affect us, and which stories contain the knowledge of a path to a good and successful life?

The Kogi know that stories hold as much opportunity as they do danger, so they attach great importance to the stories they tell, especially with regard to their task of guarding the "heart of the world." Spiritual traditions, political ideologies, religions, modern science, and also films, novels, and social media share with us effective stories about the world, which offer perspectives, opinions, and answers to the fundamental questions of life. They therefore act not only as mediators between us all and inform our reality, but also act as interpretative filters, and as such are able to either empower or weaken us. Because of the sheer capacity and influence of stories today, often ancient stories are sold to us as mere fiction, to the point where people do not believe what is being shared with them, and therefore they are not given much importance in our modern way of life.

In ancient Greece, mythos and logos were regarded as inseparable, interrelated poles of wisdom and knowledge, with "mythos" describing the figuratively intuitive approach to the world through feelings and emotions, and "logos" the logic, rational-factual view of things, which is found through mental capacity.

Through the interaction of both aspects—and only through this—a balanced knowledge and a real understanding is possible. If we devalue or neglect one of the two aspects in favor of the other, and thus turn the interaction into a conflict against each other, then a serious imbalance arises both on an individual and on a social level. By refusing to acknowledge the mystical aspects of ourselves and our world, we disregard the potential that exists beyond the purely rational mind and its limitations.

What is prevalent today among us all and remains constant, often dulled into the background with mindless entertainment and quick-fix self-help, is a longing for connection to the mystical, and the Divine, a void that is often too heavy to carry for many people.

When the balance between mythos and logos is disturbed, the result is extremism and fanaticism, among those who associate themselves with the mythical and religious side, and also among those who devote themselves to logic. Both extreme positions are gaining ground worldwide.

A simple glance at the headlines shows that religious and political extremism is dangerous. However, if we only focus on rational-materialistic science and ignore what is sacred, our felt senses, and what is beyond them, this can also be very dangerous. We are caught up in technological advancements without thinking about the consequences they might have for our existence. So just what opportunities and possibilities do we miss if we continue to neglect the informing aspects of the mythical?

The Kogi say, for example, that they know many things but have chosen not to bring them to existence from the world of thoughts (Aluna) to the material world, that is, to manifest these thoughts into matter.

Mama José Gabriel explains:

We do not speak or think badly, because otherwise we work with the thoughts of Nuanase.[9] The thoughts of Sezhankua are like the stones. They never decay and always remain. The thoughts of Nuanase are like the wood: they seem strong, but they deteriorate very easily and quickly. Our thoughts, if they are good, will not die away.

In our culture we also have myths and fairy tales that either teach us about what is deceptively enticing but has long-term devastating consequences, or something that is beneficial and meaningful. Again, we do not give these myths and stories enough attention to truly learn from them.

According to the Kogi the stories that we, the Younger Brothers and Sisters, tell ourselves and each other are "dead" stories: they do not acknowledge or incorporate the essence of life force, since we mostly create and follow unfavorable and short-term assumptions and interpretations about the world—and this is still the case even when we think we have found sustainable answers.

The questions always arise: Does a story contain and reflect the complexity of life and its fundamental interconnectedness, or does it instead become entangled in complication?

Does it include the livingness of the world or extend the reductionist and materialistic view of the world?

Does a story enhance separation or foster interconnectedness and nurturing relationships between the many different aspects of life?

Therefore, many supposed technological improvements and "sustainable" alternatives, such as the electric car, prove to be illusory models for real change when taking into consideration the actual energy and raw material consumption to produce them. We must embrace the complexity of life and look at the whole picture. However, this tendency to get lost in "hypes" and trapped in a short-term reductionist mentality can, and must, be changed. If we consciously connect to mythical stories that contain vast, powerful knowledge and we act according to them, then the age of extreme materialism and fanaticism that has arisen can also gradually begin to recede and balance can start to be restored.

The Kogi are aware that a great deal of technological development is possible, and they encourage us to use living, compelling stories as our foundation, again, both in our technology and in our organizations. A balanced and meaningful mythologic, combined with an awareness of the damaging narratives that currently guide us, integrates both these aspects into a practical wisdom that can support the way in which we live, create, and grow, so we cease constructing artificial, short-term solutions to ever-increasing, spiritual and material, long-term problems. A world in which we merge, and respect, both aspects can lead us into a future of sustainable change, founded on a holistic harmony.

"WHO AM I IN THE WORLD?" IS THE ESSENTIAL QUESTION that forms the core of our stories and builds the foundations of how we interact with the world. Modern science, which is not much older than five hundred years, answers the basic questions about life as follows: life on earth is an island within a cosmic desert of a random composition of forces created by chance, and is as such a meaningless process that merely serves the self-preservation and reproduction of genes.

Humans are the result of random genetic mutations. Human consciousness is an illusion of the brain and highly prone to error. Thoughts definitely do not influence the physical nature. Physical nature is determined by essentially dead, indifferent forces, which are technically malleable and can be manipulated, but are not suitable for communication.

In addition, different views are seen as naive, primitive, anthropomorphic, and prescientific. The history of human life on earth is a purely linear development of technical and scientific progress, which has now reached its peak in the current digital age, and this is measured by the comfort we have within our lives.

These solutions have increasingly shaped our social, economic, technological, medical, political, and historical thinking over the past few centuries. However, if we take a look at the view of the Kogi and some other Indigenous peoples, the natural state of man described by the influential seventeenth-century British philosopher Thomas Hobbes does not exist in their way of thinking.

Instead, they observe a brilliant original order, which they maintain and consciously cultivate through balance and cooperation. Furthermore, the consistent nonviolent approach of the Kogi is especially striking. How is it possible that people who have been living in, and with, nature for thousands of years and whose survival depends on studying it, to the smallest detail, and respecting it, have come to such a fundamentally different conclusion than that of a few hundred years of scientific research?

The Kogi encapsulate the question—*What is the origin of all life?*—as follows: In the living, maternal primordial darkness of being, called Jaba Sé, the consciousness of the material world is sparked into appearance, which manifests through a series of procreative processes of male and female principles (mothers and fathers of things).

The inherent law of the origin (Ley de Sé) determines the effect and the conservation of the material world, which for humans is primarily nature.

This inherent law is characterized by the principles of balance, cooperation, reciprocity (*zhigoneshi*), and by how the thoughts are ordered into territorial, individual, and communal places.

It is always about the flourishing of life (*kwalama*).[10]

Life moves rhythmically in cycles and at the same time moves forward; it can perhaps be understood in the form of a spiral.

However, the origin always remains present, and nothing artificial is added to the essence of life. The task of man is to guard and maintain this order and vitality.

So what would it look like to shift our focus from survival to a shared understanding and appreciation of the animating, vital essence of existence?

A story that answers both the questions, "Who am I?" and "Who are we?," in a way that connects us with all other living beings, and also connects us with the origin of all things, is a narrative that epitomizes and illuminates the vital essence of all life.

Another aspect that can help to characterize a "living" story is a sense of place. The stories from the Kogi, for example, leave us with a sense of the exotic and the feeling of another world. They encase a quality that cannot be put into words and are specific to the Sierra Nevada de Santa Marta. Stories that reflect a specific place give people living there a sense of home. However, home is not meant to be understood solely as a location, but as a specific way of "being in the world" and the thinking or acting that derives from that place.

Stories that are permeated with life, and as such connect us with the living interconnectedness of all things, enhance our reference points within the greater web of life, and also allow us to transcend our own desires as we become more aware of the aliveness, and sanctity, of all things. The Kogi share a dynamic mixture of stories, some of which can be understood metaphorically and others literally.

Yet, what they consistently reveal is a living world in which the Kogi are entrusted with the task of taking care of the Sierra Nevada, the heart of the world. Their stories provide a framework and a perspective of what is sacred, of how to cultivate a reciprocal relationship with nature and ensure the progress and preservation of all life.

Almost all Indigenous peoples, according to their worldview, see their territory, in some way, as being the center of the world. Many native cultures talk of a center line; just as the Sierra Nevada is the heart of the world for the Kogi, Cuzco is the navel of the world for the Incas, while the Chinese traditionally call their country the Middle Kingdom. Examples of this perspective can be found worldwide. The value Indigenous peoples attach to the place where they live as their homeland creates an emotional connection of respect, care, and appreciation for not only oneself, but also for community and the place itself.

The concept of a central point is a principle in itself; the center represents the core of life, and it is only from the center that a real deep connection can be found. Only through aligning with the center can a system be oriented, or a place, and its inherent order, be created.

The Kogi do not suggest that only their ancestral homeland is sacred, but it is the part of the sacred planet they have been entrusted to be guardians of. To truly value and appreciate both our inner and outer home and central point does not necessarily mean we must create borders and defend ourselves from neighboring peoples. The Kogi have been living peacefully with the Arhuaco, the Wiwa, and the Kankuamo for thousands of years.

In modern society too, there has been a huge rise in the trend of creating emotional bonds through stories, however, often these stories have a manipulative edge or intention, such as in the form of agenda-driven "storytelling"; companies create highly emotionally charged "stories" for their products, friends post evocative videos of their travels on social media.

In the business world, as well as in the personal development industry, people are often told that they need to be clear about why they are creating something, or offering something, and that the "why" must be woven through their entire business structure or support mechanism in a cohesive and powerful way. However, most of these stories, or "whys," are not what the Kogi would call "alive." They are essentially arbitrary constructs that ultimately only make sense within a self-sufficient perspective but are not about essential connection and order. The Kogi do not intend to save anyone, not even the earth, but they ensure balance and guard what is entrusted to them. A narrative based on life force does not construct meaning, but rather orders the actions in relation to the context and the consequences.

There are also numerous examples of stories that allow for a prudent integration of people and the places they inhabit. The survival of the Moken people during the December 2004 tsunami in the Indian Ocean is one such example. A nomadic sea people, they knew in advance that the tsunami was coming. In their ancestral stories, the waves of the ocean are inhabited by different spirits of the sea. One particularly great spirit, called Laboon, or "the wave that eats people," creates a wave that can devour whole islands. The myth says that before the huge wave comes to overrun the land the cicadas become silent and the sea retreats very far. On the day of the tsunami in 2004 the cicadas became suddenly silent, and the Moken noticed this and immediately understood that Laboon was approaching. So they fled to higher areas and all of them survived.

A similar situation was observed with the peoples of the Andaman Islands, who all survived without any human losses—except for one Indigenous group that was not originally from the island but came to it much later. Countries like Indonesia, Malaysia, Thailand, and Sri Lanka, all of which rely heavily on modern technology, suffered many deaths because they did not see the tsunami coming. Many people stood on the beach, simply watching the receding water, and did not make a connection between it and the subsequent tsunami. Although the PTWC, the Pacific Tsunami Warning Center in Hawai'i, already knew eight minutes after the quake that there would very likely be a tsunami, communication with the local authorities did not work properly, and it was not possible to inform the people early enough.[11]

Even if honing such a connection to the earth seems a distant prospect for us, stories that inform our way of living based on life force are readily available, both in a general sense and in one that is nature focused, as well as ones we can collect as local- and place-relevant traditions and legends that offer important indicators of how best to orient ourselves within the world and our own lives.

The concept of gathering old stories, myths, and legends in order to retell them, or even giving time and importance to the traditional stories of the Kogi, could elicit the impression of a desire to return to a time long since passed, that to move forward we must always look back. However, this assumption fails to recognize that the fundamental characteristic of a living story is that it contains an innate essence of the origin within it, and it is the origin within everything that remains relevant and valid, and continually adapts to inform the present.

The origin is not to be confused with the temporal beginning or the start of an event, or even the birth of the earth as such, but rather it represents the timeless essence that is active at the heart of things. The origin is therefore beyond linear time and thus always present. The river can only exist in conjunction, and simultaneously, with its source. In essence, it is about a quality one can access, not about looking backward and considering things as simply better because they are ancient.

For stories to offer truly liberating and regenerative perspectives they must acknowledge the fundamental qualities and principles that are held within them. The One Thought is a great example of this. The origin, and essence, is held within everything. It is of the utmost importance to not forget this, or we become trapped in a repetitive cycle, adhering to dead stories of limited and fabricated impact that ultimately lead us nowhere.

CHAPTER FOUR

What Did You Plant? What Did You Harvest?

During my stay with the Kogi I was often asked: "Why do you take more than you need for yourself and your family?"

Santiago, an eighteen-year-old Kogi, once asked me, unexpectedly, while we were sitting and chatting by the river in the evening sun, "When is it enough for you?"

I replied, "I don't understand exactly what you mean."

He said, "When do you feel that you have enough . . . to just sit . . . by the river and listen to nature? Or do you *always* think that you have too little?"

I told him a bit about myself, what my dreams were and

what I wanted to achieve in order to have a feeling of having enough.

He then continued and explained, "I don't just mean you, Lucas, but I want to know what you, Younger Brothers, think about it as a whole."

I responded, "What makes you think we have only one opinion as a whole about this?"

He said, somewhat amused, "Because you think alike. We do not call you the Younger Brothers for nothing, although you speak different languages and live in different territories. Here in Colombia they are building mines, and you said that this also happens in your country. Here the Younger Brothers want to buy and sell land as they also do elsewhere. You all use money and always want to earn more. That's why we say you think alike."

I fell silent. The young Kogi at my side had understood the concept of a globalized world and globalized thought. Up until this point, I had thought it was due to our ignorance that the Kogi refer to all non-Indigenous people as Younger Brothers and Sisters, but he was correct in his understanding that we have standardized our views of the world and therefore also our beliefs. Santiago and I had remained silent for a while, watching the mountains slowly fade beneath the darkening arrival of dusk.

He then said softly, "What it is all about is so simple that you overlook it, and yet it is always right in front of you. It is all there."

The Kogi are thankful for every breath, every sip of water, and every potato, and in return they offer something back to the earth in thought or material form to maintain a reciprocal balance. This relationship is shaped from a completely different awareness and way of connection. For us, it is normal to feel a general sense of lack and a desire to always want to attain more. Even if we are grateful, who among us can honestly say they have given time to be grateful for the air they breathe?

For the Kogi, gratitude is a normalcy of life, something that is second nature, and that is woven within their whole cultural paradigm, especially in the form of rituals, and even into the most seemingly mundane of tasks. It is never forced or indulged, it is simply the most natural way of experiencing life. It is something that we too can integrate back into our lives.

ONE MORNING MY PARTNER AND I FOLLOWED THE NARROW path toward the river and slowly ventured farther away from the huts of the village. The path led us through the gardens of the Kogi, which were full of lush fruits. The scent of oranges, bananas, and lemons mixed with the clear, fresh morning air, and the first rays of sunshine illuminated the colorful nature around us. The humming of insects filled our ears, and the birds were chirping; we felt as if we were in paradise. High mountains framed the horizon in all directions, and there was not a single cloud in the sky. It had not rained for months, and it seemed certain that would continue. In contrast to the with-ered yellow landscape farther down in the valley, cultivated by Colombian farmers, we were amazed by the luscious green of the plants here despite the continuing drought. There was a rustling in the bushes to our right, and a few free-range chick-ens ran past us, persistently pecking and scratching. In the distance a mule stood stoically in the shade tied to a tree, while some little pigs grunted happily over the fallen fruits, and all around us grew the small coca bushes of the Kogi, with their striking bright green leaves. The surrounding mountains were partially covered with dense forest or with a grassy shrub.

The path curved around a small bend and from it opened up a view of a group of flat stones: some were much larger than others, and they were bordered by a few orange, mango, and avocado trees. On the stones sat six or seven Mamos in their snow-white robes and their pointed hats, some with their hats resting beside them. They were all concentrating on a small wooden bowl filled with water. The Mamos held a small object in their hands and were circling it above the bowl. As we continued walking along the path toward the river, the Mamos, who were several meters away, lifted their eyes and looked straight toward us, their eyes piercing the silence as nobody said a word.

The Mamos simply sat there and gazed straight at us, and they emanated complete presence. Our wordless encounter felt so profound that it took me a moment to attune to their dignified presence. We simply stood facing each other as human beings. This moment was in complete contrast to what I normally experienced in Europe, or even the Colombian cities and towns.

Standing there in my swimming shorts, holding my biodegradable shampoo in one hand and with a towel stuck under the other arm, I felt quite the alien. Had we entered a place where we should not be? Were the Mamos at work right now? What did our presence mean at that moment? I felt the desire to stop there, to get closer and learn about what they were doing, to ask a lot of questions, and to hopefully have them all answered. Yet, at the same time, I had a strong feeling that this would not be appropriate. I knew that the Mamos were in the process of concentrating and consulting the Oracle; in short, they were absorbed in their spiritual work.

So I simply greeted them with a nod and a smile and we continued our walk along the path, which now turned right and descended toward the river. After a few minutes we reached a steep bank, from where a path led farther down to a passage through the river. A few meters upstream was a natural pool with shoulder-deep, gently flowing water enclosed by large stones. The day before, the boy that Mama José Gabriel had sent with us on our walk showed us that this would be where we would bathe.

The water was freezing cold. I felt like I was jumping directly into the water of a melting glacier higher up, but the initial shock to the system was well worth it, and the bath was deeply cleansing and refreshing. I had never bathed in or drank water that was so clear and alive. Even in the Alps I had never experienced anything like this. It was this exact same water that the Mamos, who were sitting on the stones farther up the river, were using in their wooden bowls to establish a connection with Aluna through the Oracle. The Kogi say that water connects everything that exists, and that is why they use it as a medium to enter communication with the origin, in order to receive answers for their people through the Zhatukwa, the Oracle.

MAMA RAMON GIL BARROS SPEAKS:

Mama means heat. *Ma* derives from fire, from the heat. So *ma ma* is twice as hot. For example, *mamatishé* is a thing that contains a lot of heat,[1] and the sun we call Mamayuisa.[2] Mama is like the sun. The sun never says: I won't warm this snake, I won't warm this murderer. The sun never says that; it always gives the murderer, the snake, the jaguar its warmth, no matter who it is. Therefore, a Mamo is also like the sun. Caution is advised when killing a snake, because a true Mamo doesn't kill a snake or a jaguar. A Mamo also doesn't kill trees, he doesn't kill at all. He is committed to what it means to be a Mamo; it is very sacred, very strong, and very profound.

The Mamo does not pressure the village Comisarios [the traditional authorities of the Kogi villages] by saying: "You have to do this." Or tell anybody: "You have to do that." Or order the community about: "You all have to do this." No, he never does that. A Mamo awaits the decisions of the community and authorities. A village Comisario has to hold things together, that's why he has that role. That's how the old people speak, they talk and talk and talk. Then they gather the community. The community says, "We are thinking of harvesting coffee or pulling the weeds or planting or sowing."

Then they say, "There is no coffee yet and we will not plant yet. When you, the elders, tell us that the time is right, then we will gather. We think it is best to do it like this, because in two or three months the coffee harvest will be due, or we will plant too late."

At which point the Comisarios of the village reply, "We should ask the Mamo in order to know what he thinks about it."

Then the Ñikuma-Mamo [less-educated Mamo] approaches the Mamo. There in the hut they speak and meet in the One Thought. Could it be that it's not time to harvest yet? Could it be time for a *pagamento*? Could it be that we have to communicate all together because we aim to create a new path or accomplish something else as a community?

The Mamo listens all night long in this way. In the morning the Mamos place the Oracle Zhatukwa on the hill and explore the situation. The Mamo is then able to tell whether we will meet within twelve or fourteen days or whether someone in the community has not yet attended to their own inner order. In this case we give them some more time to do this, and when everyone is ready, then we'll meet. This is how we enact the remembrance.

For us, our origin and our future are derived from maintaining the memory of the principles of life. This awareness exists in the knowledge of the oldest Mamos. If we lose our capacity to re-member, it is almost the same as if the planet would lose the sun: everything would be in danger of dissolving into darkness. The only way a human community without memory can exist, and without someone to be the bearer and guardian of this knowl-edge, would be to invent new artificial laws.

But invented memories cannot work because they are based on the will of people and not on the origin of all things. They only work in an illusory world. In nature there are no differences, there is no good and no bad. There are forces at play that are kept in balance. If one force dominates another, then chaos begins. The Younger Brothers have lost their memories and are therefore always com-pelled to invent new laws, a new law simply replaces an old law. This is due to the conflicting thoughts the Younger Brothers have.

Because of this they have power over other people but they also have wars and many diseases. Every new law implicates the need for additional laws and so on. Memory should not just be written somewhere; we live it, every day. If memory is not put into practice, it is lost. It was not made to be stored, but to be lived and shared. For us, the memory is a bit like the eyes, which were cre-ated in order to see. When we close them, everything gets dark. Only those who still have a living memory still know where they are going. Our Mamos keep the memory alive.

MEMORY IS CRUCIAL, NOT NOSTALGIC

The Mamos and Sakas are the living memory of the Kogi. In this context, living memory implies that all knowledge is important for daily life; there is no knowledge that does not serve life. The Kogi do not struggle to implement their thoughts and ideas; for them there is no difference between theory and practice. One observation that struck me again and again was that the Kogi show no sign of inner resistance. I never felt that they struggle to motivate themselves.

How is it that so often we know what would be good for us and what would support us, yet we struggle to get up and do it? In the modern world, there seems to be an increasing disconnect between human knowledge—that there is a clear and present urgency to act—and actual human action. What does this tell us about our detachment from the flow of universal energy and our essence? What impact does the burnout and stress of the majority of people have on the world? What impact does the fact that, despite billions of people working themselves daily to the bone, we still find ourselves unable to take lasting and meaningful action? The cries of "we must act now" will continue to fall on deaf ears if this undernourished and disconnected systemic way of working continues.

For the Kogi, their knowledge does not sit outside themselves. Since they have no written tradition, they must store that knowledge in their very being. According to the Kogi tradition, what is remembered, and becomes memory, cannot and must not be written down, so that living and experienced knowledge does not turn into dead knowledge. Nowadays, the Kogi have started to use audio devices to record the speeches of the Mamos, as they say that the great wisdom can still be guarded in this way, yet it differs from being written down. It is ever more imperative that the ancient knowledge is understood and embraced by a bigger part of the community, to strengthen the culture.

It is not for us to imitate their practices, because the Kogi also acknowledge that the written form of preserving knowledge is the way of the Younger Brothers and Sisters. However, it would serve us well to order and clear our thoughts so that we may remember and see from within that which is essential.

Whenever a Mamo or Saka are asked for guidance, they pose the following question: "What is the essence of this specific problem or situation?"

This question guides a person who is searching for support through a process of inner understanding and growth that leads to greater integrity and a more meaningful and resilient solution. The Kogi trust that ordering thoughts in a naturally occurring way within oneself will lead to a unique and individually applicable resolution.

For the Kogi, memory is always the remembrance of the origin. If we do not remember the origin we get lost in derivative thought constructs. The Kogi say that at the beginning of time, these original thoughts were given to all the different peoples of the earth and can therefore be remembered. So the memory is not about historical data and history in a conventional sense, but about recognizing the profound core principles of life.

FOCUS ON ONE'S OWN ROLE,
WHAT ONE HAS PLANTED

A characteristic of the original thoughts is the understanding of the One Thought, which states that in the origin there is no conflict and that all things have their unique place and inherent responsibility. The Mamos and Sakas know that there is nothing to be fought over. Rather, it is about recognizing the essential nature of things and, in case they got placed where they did not belong, restoring them to their original place where they came from. This includes the discernment between one's own affairs and those of others, which are not to be interfered with. The Kogi say that by focusing on one's own way of being and not on someone else's way, one saves one's self from creating debilitating entanglements.

Mama José Gabriel explains:

We Mamos never get sick because we don't criticize anyone. If someone talks badly about somebody else and thinks that they are not good or that they are not our brother, one can get very sick. We must neither speak badly about animals nor kill them for no reason. Sometimes when snakes come, we kill them, but if this happens we clear their thoughts seven times and order them, otherwise the mother of the snakes will come and demand reparation for her child that we killed. Sezhankua has taught us, and also you, how to do this. We still remember. We have to live in agreement with the plants, with the animals, with the lakes, with the elders, with the snow. We must also live and think in agreement with the Younger Brother. Just as we are both sitting here and talking, this is how we can live well. All of this you will write in the book.

I AM SITTING IN THE SHADE OF A LARGE TREE BESIDE THE huts. The midday sun is burning hot, and a gentle, warm breeze causes the leaves above me to rustle. Across from me sits Mama Bernardo Simungama Mamatacan and beside him his grandson Juan Carlos Mamatacan. These are not their birth names but the names they use to introduce themselves to people who are not part of the Kogi. The traditional Kogi hat Mama Bernardo Mamatacan is wearing is casting shadows on his face, revealing the many wrinkles that tell a story about his age. He is 104 years old. Although he does not walk fast anymore, he does still walk. He still has good eyesight and hearing, and most of all he still remembers his Mamo education from when he was a young boy. In the Kogi tradition, this select form of education means living the first years of life in a cave.

In Tibetan Buddhist traditions, young initiates who are recognized as reincarnated spiritual leaders, such as the Dalai Lama, undergo isolated training to develop their spiritual wisdom and leadership qualities. Similarly, the Kogi's darkness initiation for their Mamos can be seen as a form of intense spiritual training, aimed at nurturing a deep connection with the spiritual realm and the natural world.

In the 2012 film *Aluna*, a Kogi Mamo describes his experience of living in darkness as he trained to become a Mamo:

In the darkness there are no distractions. Thought is concentrated. I was taught to stare at my feet to concentrate and connect with Aluna. I was not allowed to look around or I would lose the thread. I was kept in the darkness from the age of seven months. I stayed inside until I became a man. That is where I learned to work as a Mama, to make payments for all the crops and other things too. When I came out of the darkness, the whole world was white. I stood there, staring at nature. I saw everything. The sun, the trees, the creatures. It all looked strange. The sun and everything else looked strange.

Through the darkness we can learn to see beyond the captivating allure of the visual world and experience the essence and spiritual core of existence. We can begin to merge with the intangible and unseen realm. A lengthy time spent in darkness can deeply anchor our connection to the invisible and subtle world of energy, and being devoid of distractions allows us to explore with focus and accuracy the expansion of our senses. Dark retreats are becoming more popular and are a short-form and somewhat similar way of tapping into the experience the Kogi Mamos have. Learning in this way is the absolute antithesis to the purely materialistic training we receive in school. Experiencing the dark and experiencing energies are not, for the Kogi, something to believe in or a process to achieve; they are as real and as commonplace as breathing.

So I was sitting there with a man who had spent the first eighteen years of his life in darkness. Even though I had heard about the training of the Mamos before, it was a very different experience to actually sit with a Mamo and listen to his experiences.

He speaks in the language of the Kogi, Kággaba, and his grandson translates:

I was very, very young, I can't remember exactly, when the Mamos took me to the cave. I had just been born. I grew up in the cave. I thought I was the only person in the world. Sometimes the Mamos came to visit me, then I thought that there was only us, me and the other Mamos. When I was a bit older, the Mamos took me to their house, but it was always dark. During the day I never went outside. I didn't know I had a father, I didn't know I had a mother. The only thing I was concerned with all the time were my subtle perceptions. I learned how to sing to the water, I learned how to sing to the trees. I also learned how to sing when the wind was blowing too strong. This was how I learned everything, bit by bit. I was only given pure food; it was pure, very pure. I was given no salt or meat. I ate potatoes and maggots and earthworms, they are called *binokudla*. This is how I learned everything, I was always in the cave, I only ate pure food. It was always very, very dark. I never saw anyone, but sometimes I heard the Mamos speak to me from afar. There I learned everything. I was only thinking about the things that I was told. The Mamos showed me just one place where I was allowed to bathe, only there I took a bath. Above all, my only thought was about learning more and more. So the time passed until I was given a *poporro*, when I became a man around the age of eighteen, and only then did they bring me out of the cave.

A *poporro* is a hollow gourd that the Kogi men use in their ceremonial practice. They meditate by putting a mixture of saliva, lime powder, and the juice of the coca leaves from their mouths onto a wooden stick and rubbing it on the neck of the gourd, from which a firm yellowish-white layer forms over the years. Through the scraping movement of the wooden stick against the neck of the gourd, a rhythmic sound emerges, and the Kogi say this practice enables them to materialize their thoughts.

Why are we being taught in the cave? Why are we being initiated in the dark? The reason is that it is very dark in the cave, and in the darkness are the fathers and mothers of all things. Everything that is alive, the trees, the water, the lakes, the stones, has thoughts, which are their fathers and mothers, and they live in the darkness. And this is where one learns to speak with them. They explain everything to you. The father of the trees, for example, explains exactly how to treat his children, how to make the pagamento for them, and how to work with them. All the fathers and mothers of all living things live in the dark. I learned to speak to the fathers and mothers because that is where they were.

If you are interested in something you will learn it very quickly. When I first noticed the fathers and mothers of all things, I was very interested in them. I wanted to know who they are and what they do. I did not see them but they spoke to me. If you particularly like a topic while studying, you will learn it very quickly. My school was the darkness.

Nobody taught me how to talk to the fathers and mothers. The Mamos only told me to be aware and that someone would come soon and tell me things. The Mamos told me to pay attention to what the fathers and mothers of life tell me, because they are the truth. I listened to them. People who listen to the elders will become exceptional. It is the same with the fathers and mothers of all life, one has to listen carefully and hear what they say. If not, you won't be a great Mamo.

At first I didn't hear anyone, nobody came and spoke to me, but the Mamos had said that it could take a while. Then, after maybe a month or two, the fathers and mothers came and spoke. Until then I had been sitting and carefully listening into the darkness every day. I never ate meat or salt in order to hear what the fathers and mothers had to say. One can learn much faster without salt and meat. It is the same if you eat a lot of chemicals nowadays, you no longer have the strength to go for long hikes. It is similar when you eat salt while learning things. If I had eaten salt I would have heard absolutely nothing, and the fathers and mothers would not have spoken to me at all.

There were other Mamos who grew up just like me. These Mamos already understood how the fathers and mothers will speak. They knew, even before I was born, that I would go into the cave. How could they know? The fathers and mothers of all things told them. Just as we speak and understand each other here now, they already understood the language of the fathers and mothers. It was through the Oracle Zhatukwa that the fathers and mothers told the Mamos that they had to bring the child to the cave. Once the child is in the cave, it is then both the Mamos and the fathers and mothers of all life who teach the child.

My mind was full of questions: Why do they do it this way? How do they cope with the darkness? Where do they get vitamin D from? How do they not go blind? Isn't this a horrible childhood? What is the meaning of it all?

This elder Mamo, as well as the other Mamos I had met so far, did not give the impression that they had suffered from their childhood and the way they were educated, and yet I couldn't help but feel amazed by it.

The elder Mamo continued talking while his grandson translated. He had a cheerful and happy character, which even sometimes had a mischievous childlike spirit to it. During my encounters and conversations with the Kogi I was less concerned with the details and specifics of their culture and more about the feelings that our contact was awaking in me. Here I was sitting with a man who seemed happy and deeply at peace with himself. He was very old and his age was visible, but there was no sign of complaint or frustration, he seemed lively, and mentally and physically fit. He spoke with joy about his life.

I asked in more depth about his education and he told me that as a student Mamo, also called a Kwivi, Nuakwivi, or Moro, he was only permitted to leave the cave at night to undertake what he needed to, and never on his own. The sun is considered to be a Mamo who can influence the child with its thoughts; therefore, even at night when the moon shines, the Kwivi's head is shielded with a special basketwork and is protected from the light. The places where the Mamo are initiated are often of great importance in the stories and myths of the Kogi. Sometimes these places are the origin of a particular lineage, or they are places where a particular spiritual forefather has completed a difficult task. These places are areas and special gateways to another world. It is this space that serves as the university of the Kwivi, while the Mamo is his teacher. The Kogi say: "The Kwivi does not know the sea but he knows its spirit.

"He does not know the jaguar but he knows its spirit.

"He knows no trees, no stones, no mountain peaks, no sun, and no planets, but he knows their spirits, the essence of their nature, their thoughts . . .

"He establishes a relationship with them in order to discover the interconnectedness of all things."[3]

The education of the Mamos and Sakas begins early on, while already in their mother's womb, after the baby's purpose in life has been determined by the Zhatukwa. If the baby is to become a Mamo or a Saka, then the mother follows a strict diet. She does not eat salt or meat from cultivated animals. The mother's food consists of protein-rich vegetables, which are normally only prepared by the Mamos.

During her pregnancy she will often be in meditation and stays in constant dialogue with the Mamos and follows their advice. She tells them about her emotions, and shares how she spent the day, what thoughts went through her mind, and what she dreamed. All this information is carefully observed and interpreted in order to make adjustments so that the child's development unfolds in the most optimal way. It is an incredible amount of attention and appreciation given to a person not yet born!

During the first months of the baby's life, and while the baby is still being breastfed, the future Mamo stays with his mother and his Mamo teacher in the cave. The same applies for the baby Sakas. These are the only two people he or she is in contact with. Once the child is able to eat foods other than breast milk they follow a strict diet. Special nutrition is an essential part of the education of the Mamos and Sakas, as we know from other Indigenous cultures the key role nutrition plays in spiritual development. For the Kwivis everything is prepared in clay pots: worms, beetles, river crabs, mushrooms, and vegetables, in addition to various kinds of corn, pumpkin, cassava, and different kinds of beans. After nine years they can also eat the meat of wild animals, which ensures a sufficient supply of vitamin D. They do not eat anything that does not come originally from the Sierra, no bananas, sugarcane, onions, citrus fruits, and especially never any salt or spices. All food must be white: white beans, white worms, and so on.

After puberty, the Mamos and Sakas eat meat from mammals, preferably wild animals. The Kogi say that the reason for this is the fact that the Kwivis absorb the great knowledge of these animals by eating their meat. During the education of Mamos, their food is cooked in an earthenware pot, never is anything fried, deep-fried, or smoked.

The word *Kwivi* is used for both Mamos and Sakas who are in training as well as for their "school." It means "abstinent" and refers not only to moderation in terms of eating, drinking, sexual contact, or sleep, but also to the renunciation of any form of excess. One Mamo told me that these values were the virtues of the mythical ancestral Mamos. Among the Kogi, there are many enlightening stories about Mamos and their Kwivis. Some tell of important Mamos who, despite being great wisdom keepers, made many mistakes in their lives, while others share about those who respected the values of a great teacher. Yet, the overriding theme of each story is that Mamos are still judged by their ability to find solutions in difficult situations.

Not all Mamos grow up in caves, some live in a special *nuhué*, a ceremonial house. Nowadays, some are even educated outside in the light, connected to a specific sacred place. They spend the majority of their time there and are taught directly by that place, being guided by their Mamo teacher, who are often their fathers. But still today, some young Kogi children grow up in the caves high in the mountains and are taught by the Great Mother herself.

Mama José Gabriel speaks about the process of becoming a Mamo:

A few months before the child is born, the Mamos gather and consult the Zhatukwa Oracle to see whether the child is destined to become a Mamo or a Saka. A group of maybe four women will gather. The Mamos consult the Zhatukwa and predict the gender of the child and whether the child should become a Mamo or not. If it will become a Mamo or a Saka, the mother will give birth in a sacred place. Traditionally, we conduct a "baptism" that lasts for nine days. There the elder Mamos will join and consult the Zhatukwa. They do their work so that the child becomes a good Mamo or Saka, so that he or she will become intelligent. But if it turns out that the destiny of the unborn child is not to be a Mamo, then they will be educated in the nuhué. Not many children will be taught to be a Mamo or a Saka, at most one or two per village. Those that are, learn very quickly.

The women are taught in the Kwivis.[4] This is how we know exactly who will be a Mamo and who will be a Saka. The only sign of who is a Saka is their necklaces. During the time they are still in the process of learning, they have only a few of the red necklaces, and when they wear many of them, then they are Sakas. The same applies to my hat; I also only got it after many years. When I was still in the process of learning, I was not yet allowed to wear the hat. There are even some Mamos who have finished their education but have not yet learned everything well, and despite already owning the hat they are not allowed to wear it yet. Only when they know everything can they wear it, until then the hat is set aside as a sign that they are still in the process of learning.

There is a clear priority and vast significance for teaching and guiding the young in Kogi tradition. I find it almost impossible to imagine the purity of mind, presence, and sensibilities that emerge from spending, without interruption, eighteen years in darkness. It is the Mamos belief that every encounter we have with another living being affects us, as well as the other being. The Kogi do not practice internal development and internal order simply for themselves, but for all of life. Through living this way they experience a natural balance within the universe. They live with the understanding that whatever one does for another, the benefit is received by us all.

During the first nine years of education in the cave or the nuhué, the children learn many dances, which at first are only accompanied by a subtle humming. The Kwivis dance night after night for hours on end. Movement is the foundation of life, and through dance the children learn about the structure of the world and the principles of life. As time goes on, many songs are shared that have been composed in an ancient language that only the Mamos and Sakas understand, and which the Kogi say is the language of their ancestors, the Tayrona.

Everything is repeated over and over again, the dances, the stories, everything. Occasionally, explanations and advice on how best to deepen their understanding and aptitude are shared.

During the second nine years, the future Mamos and Sakas learn how to work with the Zhatukwa Oracle and the sacred objects, how to make the pagamento, and how to understand the cycles of life.

CULTIVATING INTUITION

At the start of learning how to build a relationship with the Oracle they begin with simple yes-no questions in order to later decipher a more established, complex connection. To achieve this deepened connection, it is essential for the Kwivis to have gained experiential knowledge of all stories, and to have acquired a proficiency for listening to their inner voice. Unlike us, intuition is given great attention and is cultivated and honed with considerable specificity by the Kogi. All sacred objects are also known as *sewás*, and acquiring one gives that person authorization to engage in the specific activity related to that sewá. Sewás are gained over years, sometimes they are given by the Mamos, but they are more often directly received from the fathers and mothers of all things, the grand spirits of a particular being, element, or organism. Examples of this include permission to chew coca, to eat certain types of meat, or to perform special ceremonies.

When the Kwivis start to perform their own rituals independently, this is seen as a sign by the Mamos and Sakas that soon their training will be at an end, and that the Kwivi is now capable of working with energies individually. A prerequisite for this is that they will have acquired an extensive vocabulary, being able to name, without failure, all living beings, things, ideas, and events. This is an essential piece of work for a Mamo or a Saka.

For the Kogi it is not about what is right or wrong, good or bad, but that what they speak of is correct in terms of the original order of all things. The Kwivis also learn about social interaction during their training; it is taught in a similar way in which we understand it and consider it important, but with special attention given to ensuring that the future Mamo or Saka does not become entangled with inappropriate thoughts.

NOURISHING RELATIONSHIPS WITH PEOPLE, WITH TREES

Perhaps the most important insight that the Kogi impart is that everything consists of an actively nourishing relationship. Everything is interconnected, and when there is an imbalance, there is a debt that must be served to restore balance.

To ensure long-lasting nourishing relationships, imbalances are settled through the act of a pagamento or, in the language of the Kogi, a *zabiji*.[5] The Kogi believe that our thoughts and feelings also have an impact on nature, and the territory around us, and therefore, in the same way we restore balance with people, we must also reestablish balance in nature.

What has been given to us by nature, or the cosmos, is not "free"; by taking it we are required to offer compensation, just the same as if we were offered a service. It is important to understand that the Kogi way of thinking differs in one major way from ours; specifically, that nature is seen *as a being* with its own consciousness and motives and is therefore seen as a kin that is eager for exchange. We are truly in relation with it.

In other words, if I were a gardener my mindset would be to water my apple tree because I want it to grow so I can harvest its apples, or I could also consciously give the water to the tree as an exchange for the apples it produces. The aim here is to find out what the appropriate "currency" and "price" is in nature for the various trading partners and goods it offers. "Prices" and "currencies" are mainly based on the specific qualities of who, or what, you are in relation with and their sense of what is perceived as balance and true compensation.

Let us for a moment stretch our comprehension and experiment with an idea; assume that nature is just like us. How would I feel if my employer only gave me what they were willing to pay, and the amount was not enough for me to keep my life in balance, and my needs were totally disregarded, or vice versa? In this case, a mutually beneficial exchange for both parties would never take place, the result of which would bring a severe lack of stability and a loss of contentment and joy for life. We do not feel nourished. It is obvious that we would never choose such an undesirable situation at work, nevertheless, we treat nature in this exact way.

This type of exchange and balancing takes place on both a material and immaterial level, because there is a need for both. For example, a person needs food and water as well as recognition and appreciation. The Kogi gain an awareness of the "prices" and "currencies" in nature through their contact with the mothers and fathers of all things, and it is through this connection that they give balance for what they take. This results in a reciprocal relationship of mutual nourishment and benefit of both nature and human beings.

JUAN MAMATACAN SAT NEXT TO ME ON A PLASTIC CHAIR, ONE of only a handful of modern items used by the Kogi, sold to them by the Colombian merchants from the outside world, and began to talk:

We not only consume food and water every day, but we also kill many small animals, for example, ants. This leads to imbalances in nature. We compensate for these imbalances by making our pagamento. For example, we offer quartz crystals as an exchange and as an offering for the water we use. But we also use and consume many different types of things, such as the darkness. Also for this we have to provide a balance, because everything is alive. Also when building a house, we take many things from nature to build with, and for that we must also restore balance. We work hard to keep everything in balance. Then the houses will be preserved for a long time. If we don't establish a reciprocity, then the inequality will get bigger and from this we can even get sick.

If we do not settle an imbalance there will be consequences. In this case it is often the children who get sick, not the ones who caused the imbalance. The earth does not make a direct distinction.

When we build a house, we must make a pagamento both before we start and also when the house is finished. The first pagamento is to settle our debt to nature, the second pagamento rejuvenates the spirit of the material the house is made from. Because of this we then do not live in a dead house, but in a house that is alive. The Mamos always make their pagamento in the places of the origin of things. There are sacred places for the animals and the birds, but also for diseases, and for this we also make an offering.

When someone thinks negatively, the Mamos take these thoughts and bring them to the place of these specific thoughts. If someone thinks positively, then they are brought to the place of the positive thoughts. All of this is balance.

It could be easy to think that the reason the Kogi perform pagamento is fear of a possible disaster, but it is in fact the contrary. Their underlying focus, and the principle they uphold, is of enhancing life force. This is based on the view that each person has a natural, innate ability to restore any imbalance, because each of us holds the specific feelings, thoughts, ideas, and gifts that will enable us to create a mutually beneficial balance.

Mama José Gabriel speaks about the law of offering:

Sezhankua told us how to make a pagamento in a physical form and also in a spiritual way. We sit with the Zhatukwa and ask what material we shall use and where it must come from. The material can come from the lake or from the mountain; the Zhatukwa tells us everything. We cannot just take whatever we please. One day we might balance in a spiritual way, and another day it might be in a physical form. We have to determine all of this through the Zhatukwa. It does not always have to be one after the other, it may be that we give back by making a pagamento twice in a physical form and twice spiritually. The material we use may also change. We know exactly when to do our work, even the exact time of the day makes a difference and is important; the same applies for the duration. Sometimes we work for hours on end, other times we don't begin our work before midnight. Once this has all been determined, we then make the pagamento. These things are very, very important. That's how we live.

The material we use for the pagamento also depends on the climate. Likewise it is important from which part of our body we take the thoughts for the pagamento. When working in the mountain peaks, we take the thoughts from the upper part of the body. When working at the foot of the mountains, we take the thoughts from the lower part of our body. But everything can change depending on what is needed. Everything you need to know you can learn by consulting the Oracle. The Oracle can tell you everything—where to do which pagamento at what time. Everything! There are nine different thoughts of Sezhankua. We preserve these thoughts.

The nature of the nine thoughts were not revealed to me.

When we work with the water, we use crystals. When we work with wood, we use wood. For the earth and the thoughts we use quartz. For Kasauki [a mythical figure who cares for the forest], we work with wood. With Hayo [coca] and other plants we also work with wood. But it is the Oracle that decides all of this. We don't work that often with stones because they don't grow quickly, but sometimes we do work with them. When we work with the wind, we also use crystals. For everything we must consult the Oracle, for everything. It is not just our own decision; by working with the sewá [the sacred objects] we are guided. The sewá comes from the principles of the origin.

You will regain your own way of working, which Sezhankua has entrusted to you, and you will slowly nurture that knowledge. When it has matured, then you will do good work again. Thoughts and knowledge grow in the same way as humans do. When we are born, we are very small, we cannot walk or work, but then we grow and learn to use our body. Later we begin to walk and work well. That's exactly how we do the things. We'll all learn, but it takes time for the thoughts to grow. When we think well, Sezhankua is by our side, but if we kill someone or speak badly, he is far away.

So when the Kogi implement a pagamento and feed nature, they become co-creators of life and its processes. These are not rigid processes but ones that follow specific patterns, and, according to the Kogi, a pagamento plays an essential role in their preservation. For instance, they sustain a network of energy connections that are strengthened by a fluent exchange between two connecting points. Shells are brought from the beaches of the sea to certain mountain lakes and then stones from the shores of these mountain lakes are taken down to the sea.

However, this approach is a direct result of having cultivated a relationship of reciprocity with life force, with nature, and with the realm of spirit; it is not an approach based on a set of rigid processes and reductionist logic.

An important collective pagamento for the Kogi is the nine-day dances, which take place at different times during the year. During this time, the Kogi dance for nature, for the world, for the cosmos, and also for us Younger Brothers and Sisters.

Mama José Gabriel talks about the dances:

We dance a specific dance for nine days and nine nights, and this prevents conflict among our people, but also within the world. During the dance we carry an arrow in our bag and wear shells and feathers. The dance is for the thoughts, so that there won't be any heavy rain, so there is no war, and no diseases. We dance for nine days and nine nights without sleeping. We are awake at night, and we only think positive thoughts, we do not cut any wood, and we only eat a small amount twice a day, once in the evening and once in the morning. By sunrise we have all already bathed and eaten, then we are ready for dancing.

In the afternoon we only take a very short break before we continue dancing. The moment we sit down we take the feathers and the shells off, and when we start dancing again, we put everything back on. That's how we live. We also dance for you Younger Brothers. Thereby we change the thoughts. If someone had previously wanted to kill another person, then he suddenly realizes that he doesn't want to do that after all, that this would not be an appropriate act. Instead of murder he then might just simply throw a stone or just leave it alone. This is how we restore order in our lives.

We dance for nine days and nine nights and this is how we balance the negative and the positive, but what we do not know about we cannot balance. Therefore, it is very good that you came here so that we can hear what is happening for you in your world. The wasps are the animals that watch the earth and see who acts positively and who acts negatively. They observe everything. They tell us that we humans are far too angry and argue about everything, and that is why there are a lot of wars and there are droughts. The Wayúu Indians on the Guajira peninsula don't have water anymore.[6] It doesn't help to bring water containers there, because they are immediately used and become empty. No matter how much water we bring there, it will never be enough. Now it is time we all work together.

In the beginning people won't listen much, but we will continue our work and keep it up. If we give up, we will never live in the One Thought. We dance and sing for it. There are nine songs we sing. Nine different songs for the day and nine for the night. We see that the Younger Brothers just dance however they want. If we don't dance, then it won't rain; if we don't dance, there's no wind. Since the beginning of time we are doing this, this is how Sezhankua has taught us. Only the Mamos and the Sakas dance, the others do not, and we don't just dance wherever we want; we only dance in the sacred places where we are permitted to dance. Those of us who don't dance contribute by thinking and speaking well. During this time no one leaves the village or goes anywhere, and uniting in this way gives us strength. During these days we don't kill any animals, our focus is on sitting together, thinking and talking.

Mama Bernardo Mascote Zarabata shares:

There are different masks for different living beings. The masks help the dancers to concentrate. This extra concentrated strength makes the pagamento very powerful. All the energy generated from the dances and the masks is gathered up by the Mamos and taken to the place where it is needed. This pagamento is made for the whole planet. We work here at the foot of the Sierra in our nuhués, but we work for the whole earth. With this sacred work we regulate the extreme rains and strong storms that have been exiled to realms beyond our planet. With our work we nourish this place, and by making the pagamento we strengthen the barrier so it does not open, thus preventing these types of phenomena from entering into our world.

The Mother understood that if she lets these things come to earth, then we would face big problems. Nowadays, unfortunately, it is the case that the earth has less strength, and she is less able to hold these energies in the places where they actually belong. The extreme changes in climate that we are witnessing in recent years are due to the fact that this barrier has become weaker and more permeable and that these challenging energies are already entering our world. In the Sierra the Mamos make pagamentos at midnight to restore. While it is midnight here, it is noon somewhere else in the world and vice versa. This understanding is very important for the equilibrium. Therefore, we always make the pagamento twice, once in the day and once at night. Only in this way can we ensure that our work reaches around the whole planet.[7]

When we work, we are not concerned with our own well-being, but always with the well-being of the whole earth. There is a place here in the Sierra where all the strength of the world flows through. There we work for the strength of the whole earth. This is not the only place in the world, there are other places that are as important. They too must not be destroyed under any circumstances but must be carefully taken care of. If we do not do this, other forces will continue to invade our world and cause imbalances, and we will suffer even more drought, heavy rains, and earthquakes.

WHY DO ORACLES EXIST, AND WHAT ANSWERS DO PEOPLE SEEK from them? The Zhatukwa, the name given to the Kogi Oracle, provides scrutiny on the views and assessments of an existing situation for the Kogi. Consulting the Zhatukwa can often result in profound conversations with the earth and the universe. When working with the small wooden bowl filled with water, a hollow stone tube is dropped into the water. The movements of the resulting air bubbles answer the question from a set of structured interpretations (how many bubbles, their duration, direction of rotation, and such). The Oracle offers simple yes-no answers, but also supports complex conversations.

In their everyday life, the Mamos and Sakas pay great attention to their feelings, intuition, visions, dreams, and inspirations and then examine them in their work with their ever-present companion, the Zhatukwa. For the Kogi it is not about an attempt, based on fear, to protect oneself against the dangers of life, but about balancing energy, self-examination, and guidance.

Consulting the Zhatukwa establishes a connection to a perspective beyond the limitations and distortions of one's own perception. The Mamos and Sakas are aware that people perceive the world through certain filters, with expectations, hopes, wants, and fears, which obscure a clear view of things. By calling on the Zhatukwa, the disturbance from these filters decreases. The Kogi also pay attention to any strong bias or personal involvement they may have, for example, if they or their relatives are affected by the situation. In these instances they would delegate the inquiry to someone else, in order to receive a clear and unbiased answer.

Mama Bernardo Simungama Mamatacan explains how the Mamos and Sakas work with the Oracle Zhatukwa:

Teyuna created the Zhatukwa. He taught us Mamos how to interpret it.[8] The fathers and mothers of all things already teach the children how to communicate with them through the Zhatukwa. I learned all of this in the cave. If the bubbles in the water rise to the rim of the bowl, the answer is positive, but if they burst before, the answer is negative.

We consult the Zhatukwa whenever we need it. For this work we sit on the hills. We cannot ask questions for ourselves; in this case we better ask the other Mamos, but we can work for others.

The Zhatukwa also tells us in which month to do specific spiritual work. When to do this, when to do that.

The Zhatukwa itself is the darkness. It is the language of the fathers and mothers. There are different stones for different questions. All are slightly different.

The Zhatukwa is a mother.

What are we asking the Oracle? It depends. For example, we ask the Zhatukwa about the severe drought that occurs here. The Zhatukwa then gives different answers.

To balance a certain situation, we do this or that, we make a pagamento here or there.

Or if someone is sick, the Zhatukwa tells us why they are sick. For example, it is because they dug in the earth where they shouldn't have dug. It is always the place that is sick, not the person. When a person has a problem, the Zhatukwa tells us the reason why this is so. The Mamos know the origins of all diseases, of all natural disasters and problems. The stories explain precisely where these things come from, how they came to be, and who thought them for the first time, because problems also arise first in thought.

All things were established in the beginning of time. However, if we ourselves invent absurd artificial laws, we put ourselves into a drought, both in our thoughts and in the world. The Zhatukwa's responses are not only yes or no; instead very complex answers can be received. If, for example, the beans do not yield a good harvest, then the Mamos can find out which places are essential for the beans and should be considered. Then they concentrate and the Zhatukwa tells them exactly where the problem lies. The Zhatukwa also tells them what is needed to solve the problem.

Mama Bernardo Mascote Zarabata adds:

The Mamos never work without the Zhatukwa, and not a single decision is made without it. There are so many different combinations. Sometimes one bubble comes, sometimes two, sometimes they spin, sometimes they stay for a long time, sometimes they burst quickly, sometimes they move independently. There are a lot of possibilities. It does not matter which Mamo consults the

Zhatukwa, the answer will always be the same. Why? The law and thought of the Mother is only one all-encompassing fundamental law, one all-encompassing thought.

The answer is always the same, even if fifty Mamos gather together. Only if one of these Mamos is not centered within himself then is it possible that there will be a different answer. However, this does not mean that the bubbles move differently, but simply that the understanding of what they mean varies. In this case another Mamo comes who has a deeper understanding of the bubbles, and he will add to what has already been said by the others.

It was early morning when Mama Bernardo told me this. We had already been sitting for some time on *la loma*, the small hill, a little away from the village, on the big stones between the coca bushes. In the center of the large stones was a small pit where some objects of the Tayrona, the ancestors of the Kogi, are kept. This is also where the Mamos put the tiny cotton balls that they use for their spiritual work. Arregoces Coronado-Zarabata sat next to me and told Mama Bernardo about a journey that would soon take him abroad to talk about Kogi coffee in Australia. Coffee originates from Africa, and is not native to South America, so is not part of Kogi culture or tradition. Most Kogi do not drink their own coffee. The Spanish brought coffee to South America sometime after the conquest.

The Kogi say that a problem always appears with its

solution, meaning that after they discovered that Westerners would come and buy coffee, they also found that they could earn money to buy back the land that had been taken from them over the last five hundred years. Therefore, they sell the coffee internationally.

The Mama listened to Arregoces's worries attentively while using his *poporro*.

Arregoces spoke for at least twenty minutes, and the Mamo did nothing but listen intently. Arregoces's wife was pregnant, and they were soon expecting the baby. The Mamo began to ask the Zhatukwa about Arregoces's situation. Concentrating, he circled his hand over the bowl and repeatedly dropped the small hollowed stone tube into the water, watching how the bubbles rose to the surface. After two or three times he nodded contentedly and took off his pointed hat; he covered the bowl with it and said, "The child will not be born until you have returned from your journey. It will be born shortly after your arrival and will be healthy." This is exactly how it was, as I was to learn later. Arregoces was relieved, and we resumed our talk with the Mamo about the Zhatukwa.

A relationship with the Oracle is one of the most important relationships for the Mamos and Sakas; it allows them to accomplish and fulfill their work as true spiritual leaders for their people, and fundamentally sets them apart from world leaders.

Mama José Gabriel emphasizes that the knowledge of an oracle has been given to all people of the earth:

If the results of the Zhatukwa do not match, then we talk about it. We ask everyone what they were thinking or whether someone was dreaming, for example, about having a bath, walking around, or being angry. Then we ask what a certain person thought while they were working. If one is afraid of the outcome of the Zhatukwa, it is precisely these bad thoughts that will distort it. If someone then says that they were dreaming about this or that and that they were angry or fearful, then we take these thoughts away from them, bring them to the sacred place where they belong, and then continue to work on our inquiry. Then the bubbles will spin well to the right or to the left side and remain in one place. Then we have done our work well.

This practice has been entrusted equally to all peoples of the world. The women, the Sakas, of course work in exactly the same way. We are always more than just one person when consulting the Zhatukwa. We are at least four Mamos or Sakas. This is how we ensure that no mistakes are made. That's why there are so many Mamos and Sakas. Sezhankua has gifted the art of inquiring to all of us, but some have forgotten.

Here too some no longer want to follow the Zhatukwa, and prefer to build houses and to own cars. They invent stories rather than follow the original ones. Sezhankua has told that when the thoughts stop working with Zhatukwa, it will be our demise. We must not forget that, but continue to work with it. How many of the Younger Brothers think as you do, Lucas? Very few, right? Therefore, Sezhankua has sent you here; it was not your own decision. Sezhankua has given these thoughts to all of us. We will again live in the One Thought, which is to truly live as human beings.

An example of interaction between dreams and the Zhatukwa is the story of my own journey to visit the Kogi, which could not have taken place without confirmation from the Oracle.

As Mama José Gabriel told us:

I have dreamed of your arrival. I was in Santa Marta but my dream took me here to my village. A very old Mamo came to me and said that a brother would come, a Younger Brother. The elder told me, "Why are you sleeping? Wake up! He is coming to visit!"

I asked him, "Who is coming? And when?"

The elder said, "You have many friends far away in the world of the Younger Brother. One of them will come from another country."

"And when?"

"He will come very soon."

I then went to see Mama José Maria, and he consulted the Zhatukwa. He said, "A Younger Brother from Germany will come and it will be a very important visit."

Therefore, I knew even before you called, that you will come and that it will be important. Then I went to Santa Marta and we talked on the phone and agreed that you would come. I already knew about it. This is how we live, we use the Zhatukwa to live our life. If the bubbles don't move, we know that someone will get sick. We then ask which person, and when we find the right one, the bubbles start moving again. After this we make a pagamento for them, and in this way everything comes into balance. When the work is complete, the bubbles move toward the right side again.

Because of this, Mama José Gabriel explained that our visit was not truly our own decision, but that Sezhankua had decided it. Our planned visit already existed in the world of thoughts, and therefore the Zhatukwa could reveal that we would come to the Sierra.

The origin of all things is in the world of thought, Aluna, and from there they manifest into form in the physical world.

CHAPTER FIVE

How Will We Pass the Knowledge On to Our Children?

The morning sun was already heavy, and the sky was a cloudless blue. I had been lightly awake since the sound of the cockerels echoed through the village. Sometime later children began to play close to the hut I was sleeping in, their loud laughter becoming my final wake-up call. As I rose out of my hammock, I could almost hear their smiles, and their joy permeated me with a lightness.

It had been a lazy morning, drifting through thought and watching the village come to life, when Mama José Gabriel came to get us and we took a walk through his garden down, again, to the river. We descended toward the water's edge on steps made of flat stones. It was the same cooling river in which we bathed ourselves twice a day.

We stood on the bank and watched as Colonos, Colombian farmers from surrounding fincas that had been built on the land of the Kankuamo people, were walking with their mules to the village of the Kogi. There they retrieved rocks from the river to build their houses. The mules carrying the rocks dug deep holes into the riverbank and turned the embankment into a muddy slope.

We watched as the men working in the river loaded one big rock after another into the rough jute bags on the backs of their mules.

After a while Mamo José Gabriel said:

They are causing an imbalance. Rocks must not be removed from this place. This creates disorder. There are laws and principles that order everything and that nature follows. Even if I asked for permission to take rocks from this place in the river, I wouldn't get it. What do they think? There is a reason why the rocks are exactly there and not in another place. It is the same with the trees growing on the riverbank. They protect the water and keep it alive. They also keep the soil in place. We must not cut them and we also must not think that we have to change nature. She is perfect because she follows the order of the origin. This order is the Ley de Sé, the law of the origin. Many laws were created in order for us to live well.

It was at the time when there was no sun, no moon, no wind, no water, no trees, and no bushes yet, when the Ley de Sé [origin, darkness] was created. The question was asked, how will we live? What are we going to do?

That is why a principle now exists. When I do something that is harmful, there is a law that says that it is harmful. Then I go to visit the Mamos. The principle is part of the wind, the water, the trees. The principle has been there since the very beginning. It guides where I am allowed to light a fire, where I can fetch wood, and where I can get the rocks from, and it also governs the clouds and decides what they can and cannot do. Here in the Sierra, the clouds float in the mountains, not down in the plains; this is in accordance with the Ley de Sé. The sacred sources of water are in the mountains, all this has been predetermined. Also our body height is determined by the Ley de Sé. We natives are short and a bit chubby, the Younger Brothers are taller and slimmer. We natives also have no light or black skin, that is reserved for the Younger Brother, that's according to the Ley de Sé.

Jate Mukuákukui, the sun, rises in the east and sets in the west, this is the way the Ley de Sé has decided it, and this is what we mean when we say that things were determined at the beginning of time. Everything has an order and follows a principle.

The Kogi pursue a hierarchy of the laws of nature, which are about as real as the law of gravity to them. The Kogi believe that different cultures and people have their own unique and diverse ways of interacting with these natural principles, yet the principle of balance is universal, even if there are differing ways to obtain balance. It is important to understand that this is not about conforming to some arbitrary natured divine will, but about a sensible approach of protecting nature in a way that sustains all beings, including the Great Mother herself. Ultimately, what the Ley de Sé, the law of the origin, teaches is how to lead a good life, in harmony and equilibrium.

Even though I understood these points relatively quickly, I kept wondering: How could I begin to integrate them into my own life? How does one even apply principles? After all, I come from a world that is governed by technology and methods. So, which method should I apply to be more productive and successful? Which approach should I use to best achieve my goals?

Let us not be fooled, this is not how principles work. They do not offer methodical steps that can be systematically implemented, for example, to increase a yield or benefit a workload. They are guides to live by that generate an equilibrium in all facets of life, that do not place a higher regard for one thing, or being, over another. For there to be true balance and an optimum harmonious approach, we must consider the totality, not the individual or merely personal. Balance is about finding a point of least tension, where there is not too much or too little, but just enough for a natural occurrence to be given space to flourish. It is not about stretching something to meet a desired end or result, but about observing and adapting. One cannot force balance but must cultivate it.

It is unsurprising that in our modern society today we find a culture full of disinterest, chronic exhaustion, and lack of initiative through a system of, at best, forced balance, but more commonly, a total disregard for balance entirely.

For the Kogi, the Ley de Sé, the order of the universe, is a key characteristic of being human, because only human beings have a choice between adhering to the one original order of the cosmos and not the various artificial and fictional orders. Nature and its living beings don't have this choice, because they are the embodiment of the Ley de Sé and therefore cannot separate themselves from it. The Kogi say that if we humans do not live according to the Ley de Sé, in the long term all life will end.

LISTENING IS THINKING

When I listened to the Mamos for the first time to collect the contents of this book, I thought I understood what they were saying. However, every time I read the words of the Mamos again, I came to a deeper understanding of what they are actually saying, and also how different it is from how we live. How different their life must be, inspired by daily ritual and in the presence of the Origin. Very few of us can comprehend this way of living.

The Sé is a spiritual "primordial soup" beyond linguistic definition; it gives birth to potent ideas, thoughts, and concepts. This can be compared, in a general sense, with the creative process, which starts with an initial idea that gradually takes on its own dynamic. They both require the ability to listen. "Listening is thinking" is a saying of the Kogi.

This is also the way the Mamos explore the Ley de Sé, namely, beginning with the subtle, immaterial level of an idea or thought and not, as we do, at its physical level. In this way they gain a practical understanding of what is at stake, as they deeply understand the immaterial essence behind it. The Ley de Sé and the principles that spring from it are therefore not rigid, but are actualized within the process.

The Kogi say that according to the Ley de Sé, nature is divided into various responsibilities, which are supported and governed by the fathers and mothers of things, that is, the female and male spiritual principles. Each place and context is governed by its own unique and specific living principle.

Mama Pedro Juan Noevita says:

When the spiritual world came into existence and everything was still dark, Jaba Sé created the origin in the form of spirit and thoughts. She was not a human being, neither was she made from air or any other physical element. She was only formed of thought. She looked around and then considered it would be right to create people. However, she couldn't arrange the whole universe alone, she needed the help of the other spiritual mothers and fathers. Therefore, she requested the presence of the spiritual mothers and fathers named Kudlula,

Shuakala, Mulkueike, Kalbasankua, and Gonawindúa in order to discuss the situation and to assign the various responsibilities to each of them and to form the spiritual order.

She defined the places from which the order should be established. She understood that it was very important to first structure and organize everything. If one does not organize oneself, then there is no center for one to speak, to think, to guide, to listen, or to function from.

Nowadays, we still remember the real names of people, animals, and nature. This is how Jaba Sé created Sezhankua, the territory, nature, the people, the animals, and the thoughts. Sé governs simply through her presence. Sé organizes, Sé thinks, Sé flows, Sé moves and lives. While creating the spiritual world, Jaba Sé conceived who rules, who is in control, who makes decisions, how far the spiritual world should reach, and how influential in scale and effect it is. After this, Jaba Sé gathered with the spiritual fathers and mothers and together they determined who would take on which responsibility, and then Jaba Sé told each of them what their task was.

RENOUNCING VIOLENCE

The Ley de Sé imparts very strict guidelines on how to behave, which for the Kogi is to live a life having fully renounced violence. They live this to its fullest extent, an extent that may seem too extreme from our modern world viewpoint, yet for the Kogi, it is an inherent part of how to live well.

That morning, when we stood with Mama José Gabriel at

the riverbank and watched the Colonos loading rock after rock onto the backs of their mules, despite the fact that no rocks should be taken from the river, we asked the Mamo what measures he proposed.

Mama José Gabriel replied:

We see that what they're doing is not right, but we must not criticize them or talk badly about them. We might think: "Well, someone did something bad and therefore I am allowed to criticize them," but that's not true. Many people think that it is all right to go down a certain way, even though it is forbidden; they simply think, what's the harm? We might think that we are only criticizing one single person, but in fact we are criticizing the whole earth. Why is that so? Because we are all part of the same life, and everything is intrinsically connected.

For example, it is not good if we look at a mountain that we do not like and think badly about it, and then turn our affection toward another one; that's not good. The mountains will ask, "Why did you look at me and think badly?" But if you look at the mountain and say, "How beautiful, there must be a sacred place there. I will respect it." Then the mountains will be grateful and say, that man is a friend and he has helped, he did no harm.

Nature has many aspects, and each aspect has some kind of governance. It is like in Colombia, there is also a president who decides everything. The same is true in nature. The mountains, the water, the rocks, they all have their own law that determines how they express themselves and behave. Everything is alive; this is how Sezhankua has created it. Even the trees, stones, and water can cry. That's why the elders here talk so much and explain how best to take care of everything.

I have also heard that you, Young Brothers, punish people when they have done wrong, but it's not about punishing people; it is simply a matter of stopping what you are doing wrong. It's hard to stop, but it is the only way. If all we do is just talk about it, nothing will change, and the Younger Brothers must play their part in it. We'll all think well together. We are still guided by the One Thought.

Put plainly, translated into terminology we can comprehend, Mama José Gabriel is saying: The best course of action is for the men at the river to simply stop their work and ask the Mamos to consult the Oracle on where best to take the rocks from. It is fundamentally about order, and to access order we must work together in harmony. The Kogi truly do not have a warrior culture.

THE RESPONSIBILITY OF THE INDIVIDUAL TASK

Once again I was sitting in front of a small, loosely thatched hut with Antonio Coronado, the father of Arregoces. We were in the village of Awiaka. Antonio was sitting on a low bench while carving a piece of wood. I had taken a seat next to him and was watching his work take shape. As is so often the case, the Kogi looked up at me briefly, and without saying a word he nodded in my direction and continued with what he was doing. With a machete as his only tool, he cut planks from approximately twelve-centimeter-thick tree trunks, removed splinters, leveled their surfaces, and then made notches at regular intervals.

It was a cloudy day, and yet still the heat persisted, the air clung to me. In the far distance I could see a world of black clouds forming. The imminent arrival of thunder felt certain. There hadn't been a single drop of rain in all the seven weeks I had lived with the Kogi, and all the plants I'd seen looked beyond parched.

I said, "Maybe it'll rain today."

Without looking up Antonio replied, "No, it'll move on."

I asked, "How do you know that? The clouds look really dark over there, it looks like a thunderstorm."

He smiled. "I don't know, but the birds do. There is a certain bird that calls before it rains. But today they are silent."

I was quiet for a while. The small sand flies were once again a real nuisance. I tried in vain to keep them at bay, while they seemed not to bother the Kogi a great deal. I asked, "What exactly are you creating with that wood?"

"It will be a frame to make panela.[1] The wooden slats will be interlocked and then fixed to a panel; after that, we'll pour the liquid sugar into it and it will dry," he replied. Holding one of the finished pieces of carved wood aloft while gently appreciating it, Antonio said, "We'll make good panela with it."

The precision-crafted slats are made according to an exact

principle and will undoubtedly fit together perfectly. All of them made with just one tool, the machete.

The Kogi have a close relationship with the activities they pursue, as well as their tools and their self-made objects. Each task requires a permit, a *sewá*, which is often officially authenticated in the form of a specific object. A young man, for example, is not allowed to marry a woman until he has passed an initiation and received a *poporro*. In Kággaba, the term *sewá* means permission, guarantee, access, awareness; when addressing someone as *na sewá*, it also means "my darling." A sewá represents a particular alliance and is like a certificate, comparable to a driving license. It is often a piece of string put on a person's wrist. It is most commonly handed out by the Mamos, and then they tell everyone that the person has it.

This permit offers a kind of protection, just as road users are protected from someone who is possibly incapable of driving through the necessity of obtaining a license. At the same time, the license also protects the driver himself from making unnecessary mistakes, such as ending up in a ditch or hitting a tree. In a similar way, certain permits must be renewed at regular intervals, checking that the necessary understanding and sufficient skills are still in place. You would be correct in thinking this all sounds familiarly bureaucratic, because it is. When I explained our modern world obsession with acquiring permissions to the Kogi, they nodded in approval, but also noted that it is imperative that we always take into account the spiritual side of a situation.

Mama Pedro Juan Noevita shares the original story of the sewá:

According to the *shibuglama* [story, wisdom], singing was born to order the material world as well as the pagamento, the *esuamas*, the people, the justice, and the territory. When they had everything ready to govern, they taught Jaba Seinekun and Jate Sezhankua how to make pagamento. Then, through constant ordering, they maintained the territory, nature, people, animals, and the thoughts. Sé-earth, Sé-thoughts, Sé-people, Sé-crystals, Sé-animals, Sé-singing, Sé of all things. When we step out of this order of all things and ignore Sé as the guiding principle, then there will be chaos. That is why it is important that we allow the sewá to govern; the sewá of the thoughts, the sewá of the pagamento, the sewá of the people, the sewá of the animals, all sewás. They are the spiritual fathers and mothers who decide how we order, how we lead, how many thoughts we carry within us, and how we reveal stories.

The bearer of the physical object attributed to a specific sewá holds knowledge of all its material and spiritual implications, and also all potential imbalances that may occur and how to use any associated materials, and what they may bring. Unlike a theoretical and practical driving license test, the applicant of a sewá is also given a spiritual understanding of the practice. In terms of road traffic, this would mean, for example, checking mental stability at rush hour and the driver's general sensibility for nature.

Just as every living being has a father and a mother, for the Kogi, every object and every practice has its origin in two poles. Because of this one must seek permission from both the father and the mother of the task. So, for example, if someone wants to plant beans, they need a sewá from the father and the mother of the bean. This assures the responsible handling of beans on all relevant levels. Some activities even require a whole catalog of individual permits, such as the construction of houses, for all the types of wood that may be used and their processing and so on. Architects who work with state building authorities will find that things are no more straightforward among the Kogi. Since cutting down trees involves severing a living tree from life, this act must be balanced and reversed on a spiritual level. This revives the house so that the Kogi from then on live in a house made of living materials. Knowledge and skills in this field belong to the group of house-building sewás.

In total, there are as many possible sewás as there are activities and materials, some sewás being reserved for only Mamos and Sakas; however, even they do not receive all existing sewás.

Aside from one exception, sewás are only given to adults, depending on gender, age, experience, and function. A Kogi reaches the age of maturity at about eighteen years old. When babies are born they receive a specific sewá, most often in relation to what food they consume. Then for many years they receive no further sewá, but when they grow up and become an adult, they will receive more sewás.

Each sewá expands the bearer's consciousness within the fabric of life, as he or she will have learned how to holistically take care of a specific task.

Mama Ramon Gil Barros explains:

Every culture in the world has its own criteria and processes for organizing thoughts and passing on traditional knowledge. Sewá derives from Sé. Newborns receive the first sewá. Later, when we receive our *poporro*, we receive our second sewá. This is the one that gives us our responsibilities, shows us the path, and serves as a guide in our relationship and communication with our energies. It is understood to be the material power that guarantees the traditional knowledge of the laws passed down to us since the origin. In order to carry out any kind of task, from cutting down a tree to exercising a traditional authority or being a Mamo, it is necessary to obtain the appropriate sewá. This is the only way we can fulfill our tasks.

To receive the sewá, it is required to complete all of the necessary work beforehand. We then will understand them all and can put them into practice, every single principle from the Ley de Sé that is relevant to it. Only then can something that is a material symbol of spiritual responsibility be handed over. Once it is received, we adhere to it. We can no longer just walk around looking for the Mamos in order to constantly ask them questions. A sewá cannot be simply seen as a material object; it is the embodiment of knowledge, of culture, and of spiritual order. Not every sewá is for everybody. With the sewá we order and fulfill the principles of the origin. If one does not accomplish this, one contradicts the Ley de Sé, and the sewá is hurt, and problems can arise from this.

The sewá can only be given to a man after he has received the *poporro*, and to a woman only after she has gone through her right of passage initiation from girl to woman and after she has received her spindle. Before this, only once, at birth is it possible to receive a sewá, because one has neither the knowledge nor the necessary wisdom or maturity.

To receive a sewá, the person must first sit and order their thoughts, then order everything that concerns their individual behavior. The sewá determines the place where the necessary task has to be done, as well as the time required for it.

We first have to acquire the required knowledge: it takes between fifteen and twenty years to become a small Mamo, thirty or forty years to become a Mamo, six years to become a traditional authority, and two years to become a leader.

The sewá itself and the *esuama* of the lineage determine from which Mamo, and from where, it shall be received. After receiving the sewá, the task that the person fulfills in the community is specified, so he or she can direct, guide, speak, teach, or lead and advise others in order to avoid problems and find solutions. Receiving a sewá does not mean that a person already has enough knowledge and skill to continue their work. The work is a constant process, the sewá is continuously renewed during the process of learning, both spiritually and physically. To receive it means to make an everlasting commitment to the laws and the behaviors that identify and define us as native people. A sewá represents and symbolizes the beginning of a task and craft that one is associated with in the community. Without this, it is impossible to achieve satisfactory results that are in harmony with the principles and our coexistence.

A sewá does not expire after a certain time or transition; implementing the traditional order into practice is a permanent responsibility. The stories and the principles provide guidance. A sewá allows us to understand the knowledge that is stored in our territory, to connect to it and to value and respect it. We also gain access to the knowledge of how to organize our community according to the laws of our ancestors and how to use the ancestral lineages to establish the required order. The Mamos process of learning is like a big piece of fabric, it is like weaving. They are constantly accumulating ancient knowledge through the guidance of the Ley de Sé.

We order our thoughts through material and physical work. In doing so, we learn about the Ley de Sé step by step. We live in deep relationship with the ancient knowledge, and that guides our individual and our collective behavior. The care we give to preserving nature, as well as the granting of sewás, is a reflection of that knowledge. The more we deepen our understanding of the Ley de Sé, the more our duties and responsibilities in the community grow, until at some point we reach the level of a traditional authority.

WE ALL ARE NECESSARY

A fundamental characteristic of a sewá is that each being has a unique and intrinsic responsibility in life. The individual task of an animal or a plant within the natural structure of life—in this case the ecosystem—often seems much clearer to us than our own unique task in life. If, for example, bees became extinct, most ecosystems would collapse within a very short time, because without bees there would be almost no pollination, and this would lead to a chain reaction of other species becoming extinct and parts of our ecosystem collapsing. Quite quickly this chain reaction would have an alarming effect on humans. In regards to some species whose contribution to the ecosystem is less obvious becoming extinct, it worryingly may not become clear what effect that causes until it is too late.

Arhuaco Mama Wintukua Kunchanawingumu shares:

Every human being is born with an individual task. We must never ask anything of them other than what their responsibility is or what is uniquely intended for them by the Ley de Sé. When you discover your task, follow it and accomplish it, then you lead yourself to health. There are certain rules about what everyone can and cannot do, but everybody is not meant to do everything. Sometimes a certain disease occurs because we do not follow our task that is intended for us from the beginning of time. One also has to know how to grow that calling and how to use it, otherwise one gets sick in one's thoughts and in one's body. That is why it is so important to know very clearly why we are in this life, where we came from, and what it is that we need to do.

The place where we are born is the most important, it is there that the seed of what we will become in our lives is planted. That is why it is so important that during the ceremony of birth we always consider what the child's task is and how they will fulfill it. If we don't do this, diseases can arise. These are the basic characteristics in our culture of how we view health. If we work with Western medicine at all, we always keep this in mind. We are not alone in the world, we are part of the world, and even the very thought that one can live alone is a disease. Therefore, it is important to heal the ancestors and the parents and grandparents as well as nature in order to heal the human being. The deepening of this knowledge is reserved for the education of the Mamos.

We natives know that we are not alone, that everything is connected and that all living beings are interconnected. Sometimes one might think that it is possible to act alone, but one can't, because everything always depends on and influences one another. That is why it is so important to maintain a relationship with the ancestors and forebears. We always offer them our respect, we cultivate their fields, and we nourish them. We live in harmony with all living beings.

According to the Kogi, the Ley de Sé has provided everyone with a specific place and a specific task in life. This means that I am only responsible for tasks that have been designated to me and not for the tasks of others, regardless if their tasks seem more enjoyable or garner more respect. Even if my own task does not seem particularly spectacular, I will find that fate will fall into place because a task is governed by the Ley de Sé.

What is meant for us feels easy because it is natural and intrinsic, while what is not meant for us can be let go. There is a saying that reads, "You can't be a jack-of-all-trades." The Kogi would never consider not adhering to the task that was given to them, and just the mere thought of wanting to pursue many different tasks all at once would be viewed as a confusion of the mind that would need to be immediately resolved with a Mamo.

With this way of thinking in mind, our widespread belief that for a good life we must compete for our position and status is both absurd and an illusion. The so-called American Dream, which suggests that one can become anything they want if they only try and work hard enough, is the complete antithesis of this fundamental Kogi principle. You cannot, and should not, become anything you want, because realizing your individual innate self is of great value and is necessary.

It is far less a matter of looking outwardly at what attracts us, and about looking inwardly at what comes naturally to us and unfolds with grace by itself. What unfolds is always in service to life. Our individual innate self is not self-contained but interconnected, and realizing our individual purpose, if fully aligned, will benefit both the earth and the individual equally. There is no such thing as something that is truly beneficial to life and not to us, and vice versa.

In our culture, it is common for people to undertake all manner of trainings and courses, often the only prerequisite being that you pay for it. People adopt a certain trend and lifestyle without seriously considering their motivation for it and without connecting to whether it resonates with their own individual, intrinsic nature. The Kogi would say that by doing this we create chaos in our lives and artificial emotions, thoughts, and actions that ultimately hinder or harm the fabric of life. By not having a wide range of free choice in what we choose to spend our time doing, this may seem like an unacceptable limitation, but for the Kogi such self-oriented behavior is a violation of the law.

The Kogi, and many other Indigenous peoples, have a completely different concept from us of what we refer to as a "calling in life." The idea that we must all follow our passion or a desired career is not on their horizon. Instead, they have the lifelong support counsel of a Mamo.

A fundamental problem with our way of living is that we assign ourselves a task rather than going inward and cultivating a relationship with our innate task. The former way often only satisfies a need to feel important rather than guides us toward establishing an essential connection to ourselves and the natural order of things. We then become governed by what society deems as important or impressive, and not by the more subtle internal voice that offers natural, innate guidance. Considering our societal propensity for diminished motivation, lack of clarity, and even depression, this way of living that

the Kogi demonstrate could be a great indicator of how we
may also look for guidance.

Later, Arregoces Coronado-Zarabata explains:

At the beginning of time, the Great Mother left her knowledge to
everybody. Therefore, when the Kankuamo lose their knowledge,
they must go to the Kogi and ask them so they can get their
knowledge back. We are the four peoples in the Sierra who do just
that, if necessary. It is similar in other places also. If a people has
forgotten something, they can go to their neighbors and ask them
for it, and if the neighboring people still has the knowledge, they
can share it. This knowledge can then facilitate access to one's
own wisdom. But they will never explain everything and never in
every detail. It's all about remembering. The remembrance of
what a people originally had themselves, what thoughts they had,
in every detail. Whenever we Kogi have explained something
about spiritual work to our neighboring peoples, it is only as a
door opener so that they can find their way back to their own
knowledge and the unique form of their path.

Mama José Gabriel continues:

When we ask ourselves how we take care of the things, these are
the thoughts of Sezhankua. We are not trying to live like you do in
Europe or North America, and you shouldn't live like we do here.
Even those of us who live here at the foot of the Sierra live differently from those who live in villages at the top of the Sierra. Even

that, Sezhankua arranged. Everything has an order. You come from another place and because of that cannot live here, just as I am not to live where you are from. I am not allowed to tell you exactly what you should do, because if it is used in your place you could cause damage there.

Some native peoples are also beginning to adopt things from the Younger Brothers. If we think badly, we will never preserve the origin. I was here in Colombia with the Muisca people and they asked us Kogi how to make the pagamento. We responded by asking if they had any idea of how to make a pagamento. They said no. We then asked them if they did not know of Sezhankua either. They said they were not sure. We then told them that they had to find their own way again and that we cannot tell them how to make the pagamento. They must not imitate us. If a people has forgotten everything, we don't tell them anything. It would only cause harm.

Sezhankua has left an individual legacy to all peoples on earth. We Kággaba must sit in the *nuhué* and think there, but that doesn't mean other peoples have to do the same.[2] If there are sacred stones, we sit there and work with the energies. For Sezhankua, this is all the same. If we forget these things or stop doing them, we will no longer live well. That is why I say with complete confidence that we all must only have one way of thinking. Then we don't argue and we work well together. That's how we live. We ask ourselves, how are we all going to think alike, since we are billions of people? But for Sezhankua we are only very few, about three or four, not much more. We will work in unison, in great unity. That is *zhigoneshi*. If we don't work in the spirit of *zhigoneshi*, then in the long run we will fail.

We all live under the same sun, we all drink the same water, we all walk the same way, and therefore we can and should also think in the same way together! If I do something a certain way, my own child will do it the same way. We ourselves teach each other everything, no one else. This we have to bear in mind and we have to take care of it. Therefore, it must be all of us who think the thoughts of the Ley de Sé again. The principles that Sezhankua left behind apply to all of life, to nature, to the Younger Brother, and to the Elder Brother. Sezhankua has not entrusted us with two different ways; he assigned us only one way.

FINDING YOUR OWN INNATE WISDOM

One's own intrinsic nature cannot be in competition or conflict with the intrinsic nature of another, but only with what is not natural. The Kogi's view is simple: if we no longer follow the Ley de Sé we create an imbalance that harms the earth, other people, and, above all, ourselves. Therefore, it doesn't make sense to transpose Kogi traditions or practices to other places in the world. The principles and practices they live by are right for them in the Sierra Nevada de Santa Marta, and our own are what will suit us best. The Kogi can act as guides and advisors on how best we can reveal again what is innate within us.

Mama José Gabriel says:

The world is in great disorder, but here in the Sierra things are still in order. We are all the children of Sezhankua. We all eat the same, we have to think alike and work in the same direction. Who can naturally be without water for five days? Can you? No, neither can we. We are all the same.

Once a man came to me and said that we natives knew much more than the Younger Brothers. But I asked him whose child he was, if he was also the child of Sezhankua, to which he confirmed he was. So, how is it possible that we do not follow the same knowledge? We speak differently, we sing differently, but in essence everything is the same. The rooster always crows at the same time; how does he know when to crow? He crows early in the morning and then at noon. The hens do not crow; why? They are children of Sezhankua, he has told them when to crow, they have not invented it for themselves. If a hen started crowing, this would not be according to the order of things, it would mean that it had left the path of Sezhankua. The roosters and the hens are like us Elder Brothers, they still remember the tasks that Sezhankua gave them. The roosters crow, the hens cluck; that is how it was entrusted to us at the beginning of time.

ANOTHER ASPECT RELATED TO THE TOPIC OF ONE'S OWN IN-trinsic nature would often come up in conversation, specifically the effects of foreign thoughts, substances, and the spread of nonnative plants within the territory of the Kogi. They say they have experienced the negative impacts of all of these, yet in some cases also some positives.

Juan Mamatacan explains:

The Colonos brought new plants that do not belong here. We then started eating these plants too. This is not good for us. If one eats food that is not originally from the place where you live and where your ancestors came from, then one is poisoning their body. Many chemicals are put into nature, herbicides also cause great damage. Today we have some diseases that come from the poisoning of nature with these chemicals. We eat these things and then we are also poisoned. All the chemically produced medicines are harming us. The more artificial chemicals we ingest, the more we become artificial and chemical ourselves. In this way, in the long run, we will lose our culture, it will change the way we think.

The government wants us to believe that vaccinations are good for us, but we have observed this very closely. They work a lot with machines, so their thinking is not aligned with the principle of life force anymore. Some Kogi who have had chemical medicine forced on them by the government are not the same as before. They no longer think well and have become slow. We Kogi work to preserve our culture and identity.

Many of the trees that grow here are not native either, the species were imported. Just as we let trees and plants grow in places where they did not originally come from, we do the same with people. Today, many people live in places that are far away from the places of their ancestors. They also mix with people who are not from their people. In doing so, we forget how to guard the places.

Juan Mamatacan's words are an important example of how artificial interference can affect the Kogi, from culture and agriculture to medicine and nutrition. Culture is the collective expression of an intrinsic tradition, namely, one's own thoughts. Therefore, when we merge food, culture, or diseases, this primarily means we confuse our thoughts and cause disorder.

So what to do when the order has been disturbed? Well, the Ley de Sé is applicable here too. As soon as the disorder has been rectified, a process of natural realignment begins. This has been seen with, for example, cultivated plants that when released back into the wild, and are no longer planted in monocultures and treated with chemicals, revert back to becoming wild species and acquire much more vitality and strength. This happens without human interference, but through the self-regulation of nature.

Mama José Gabriel explains:

When the Younger Brother arrived here, he killed many of us Indigenous people. But a few remained as seeds from which we are now growing again. The same applies to your own original thoughts. There are still many animals in the mountains. However, the Kankuamo put salt on the meat of the animals they eat.

That is why many Kankuamo die, the Mamos say. Here, the animals are eaten without salt, otherwise our thoughts are damaged.[3] We haven't all understood how to live yet, but if you have a Kwivi, a school focused on the origin of all things, you will teach it. And those who have learned the knowledge will pass it on to others. In this way, many people can be taught.

We Kogi are not allowed to breed many animals. If we raise many cattle, many chickens, or many sheep, then we step out of the Ley de Sé, but if one thinks like a Kwivi, then we will learn again. We will remember again to eat less meat and especially not industrial meat. We should also eat less salt.

Way up in the mountains, they still do not eat salt or farmed meat. One must not take dogs or mules there, because they don't belong there, it is not in the order of things for them to be there. When those who live up there do their spiritual work, they do it very well. All of us who live here on earth still know many things, but some are willing to learn, and others are not interested anymore. That is why we face great problems. The Younger Brother also still knows a lot, but it is vitally important that he does not forget it again under any circumstances, and, above all, he must apply it, otherwise we will harm the earth. There are eighty-six different native peoples in Colombia; we all speak different languages, but we work in the same way.

Three days after our arrival, having almost settled in to our new surroundings, Mama José Gabriel came up to us and explained, "You have to register, it's like in a hotel. There you have to register at the reception and you have to do the same here in our territory. You cleanse your thoughts, reveal your intentions, and come into balance with the land you are on. I gave you the last few days to arrive and to realize where you are now."

We started talking and he asked us to remember where we had come from, which problems and worries we had left behind and what their origins were. We shared our responses, and he listened carefully. He did not look at us but was concentrating on his *poporro*, rubbing the wooden stick against the small yellow gourd.

After a while he said, "You have told me various things and thoughts now, but they all come from the same essential core. The Younger Brothers twist their thoughts and turn things upside down. You often do things without knowing why you are doing them that way and without asking yourselves if it will be in balance.[4] You no longer consider the world, you no longer consider your territory, you no longer consider the place where you were born; instead, you are busy with your desires and your own will and no longer with what you have within you to give to the world."

What do you have within you that would serve the world best?

What do you do with ease?

What feels natural to you and ignites curiosity and excitement within you?

CHAPTER SIX

What Do I Have to Offer the World?

During our first few days we had time to experience the territory of the Kogi, to feel, and to slow down. Mama José Gabriel said, "The thoughts of the Kogi are connected to this land, and only if you feel and understand this place, can you then comprehend the Kogi's thoughts."

As a visitor to the Kogi one is often waiting for hours, because waiting is part of acclimatizing to the territory and is expected of a guest. I had to quickly confront my frustration that nothing much was really happening, that waiting, and not knowing for how long, was part of my arrival process. It took time, but soon I was able to let go, which gave way to a space of stillness.

People are inseparably connected to places, and places are inseparably connected to people. In the stories of the Kogi, the creation of the territory is equally as important as the creation of humans. Nature is alive, and places have thoughts that are embedded in them. The form of the world appears in many aspects of a Kogi's life. Every mountain peak is a world, and every house is also a world, inhabited by consciousness, and conveys reference points such as the center, the door, and the different levels. The Kogi say that Jaba Sé created the cosmic egg at the beginning of time and placed it between the seven points of reference: North, South, East, West, Zenith, Nadir, and the Center. These seven points are associated with countless mystical beings: animals, plants, minerals, colors, winds, principles, and values.[1] The cosmic egg is divided into nine horizontal worlds, and the fifth level is where we humans live. The levels are the nine daughters of the mother; each of them has a different color and represents a different world. Built in the funnel-shaped roof of each *nuhué* are four layers—still in the present time and since time immemorial—which symbolize the four layers of the upper worlds.

The nuhué floor is our human world but it is imagined that beneath it lie four invisible layers, within an inverted mirror image of the upper nuhué. Therefore, the center of the nuhué is also the center of the world, because it is an exact representation of the universe. Every nuhué, or house, as well as every bag, pond, or clay pot, represents a womb, and the land is not just simply the land, it is the universe and it is also the human being, just like the human being is also the land.

Mama Ramon Gil Barros talks about the structure and the order of the territory:

It was Sezhankua's responsibility to organize the world according to the Ley de Sé. First he arranged the stones, the structures that are the density and strength of the world. From there he took the thread of the thoughts that he had received from Jaba Sé and passed it through the center of the earth, through the mountain Gonawindúa, which is a mountain that reaches high up and also goes down far into the earth. This is how he brought the place of creation into being. At the three outermost points he manifested Kadukwa, Shkwa-kala, and Shendukua to preserve the physical world, and at each of its four corners he placed a guardian so that the material world would continue to follow the cycle of life's constant regeneration.

Sezhankua is the main creator, the original authority, the one who bestowed each being with its task and its function in nature. He was also responsible for the laws and the principles of harmony and coexistence.

However, there was no fertility in the world yet, and that's why Mother Seinekun appeared; she was the black woman, the fertile earth. It was she who organized everything, simply everything that exists on her.

The union of Sezhankua and Seinekun, of male and female, was the origin of many authorities of the physical world:

> Kalashé and Kalawia: the authority of
> the forest and the trees;
> Gondwashi: the air;
> Mamatungwi: the sun;[2]
> Zareymun and Zairiwmun: the sea;
> Zanani and Zarekun: pets and wildlife;
> Ulukukwi and Ulukun: the snake;
> Seaga: the tigers and pumas;
> Kakuzhikwi: the ants and so many more.

The original order and harmony are in the Ley de Sé:

Sezhankua holds the authority and organization, Seinekun the governance, the practice, and the use of our territory. The three of them, Sé, Sezhankua, and Seinekun, represent a union of our vision of how we prosper and the original order of the ancestors of the territory. Sé, Sezhankua, and Seinekun are the fundamental principles that determine our mission as human beings on this earth. However, it is not their task to maintain and organize the material world; this is our responsibility. This is the only reason why we are alive, and also the implementation of the knowledge of the Mamos has purely this one purpose.

Before the material world existed, the fathers and mothers knew how to give birth spiritually. They did not have a body, but they knew how to cultivate sexual relationships in a spiritual way. What we can observe in our territory today is what they gave birth to. We know every part of the mountains, everything has its name, every place and every thing has its unique expression. We work in various places, and through the Mamos' work the territory is cleaned. We work with the cardinal points, with the winds and the rain. We arrange the animals and the plants by doing our work in the places of their spiritual fathers and mothers. The male and female thoughts are collected and brought to us and are then offered in the appropriate place. This is how we organize and receive clarity. When we are in the process of ordering, one easily understands our territorial order. In some parts of our physical territory we are not allowed to cut down trees, and on the plains we are only allowed to cut a few trees in order to sow something there and to protect it. If plants suddenly grow in a place where they don't belong, we reestablish the order. We often talk about maintenance, and we order the territory in a physical way as well as in our thoughts, in fact, every part of it.

It was only after having had many conversations on the subject of place and territory that I realized how widely applicable the views of the Kogi are. The importance of the place of one's ancestry and territory is essential and of immense value for the Kogi. The territory contains the basis of identity for every human being and gives each of us autonomy by enabling us to achieve all our basic needs, including all necessary spiritual and mental information, without creating dependencies or obligations with other people or territories. The idea that "the grass is greener on the other side" does not exist for the Kogi, and traveling outside their own territory is not a desirable experience for most of them. Their territory is enough for them because they invigorate it.

The Kogi say that our scattered thoughts, our doubts, our uncertainty of who we truly are, and our urge to reinvent ourselves are all essentially due to our territorial dispersion. We are not completely present and immersed in our own territory. Instead, we distract ourselves with the desire for more and more, while often the only thing that truly counts is comfort.

Territorial dispersion and mental, as well as emotional, distraction are one and the same for the Kogi. We eat food that is imported from far away countries and that therefore has little or no connection to the places where we actually live.

The Kogi see that human beings are as deeply connected to their territory as to their own limbs. People who are conscious of their body know how crucial it is, for example, to be aware of numb areas, and to focus their attention on these spots and rejuvenate them. Just as metabolic disorders, tensions, and stress in certain organs may occur within the body, they also occur within the territory.

Most of us live our life secluded in spaces such as our homes, our workplace, the car, or on social media. In fact, the internet has become a substitute for our inner territory. While cultivating contacts and connections through social media, we ignore the actual land as well as our bodies, so much so that we use the TV shows we watch regularly to get a sense of familiarity and comfort when we need it. The Kogi see our place of birth and ancestry as a mother who gives birth to us, just as our biological mother did. When talking to the Kogi, we always talked about *aquí* and *allá*—here and there. The fact of whether someone or something is *de aquí* or *de allá*, from here or from there, is of utmost importance for them. I tried to explain to them that those of us living in modern urban cities are less and less "from here" or "from there," we are more and more from everywhere and therefore from nowhere in particular. The Kogi say that by being detached from a specific place, we become detached from the order of thoughts until we can no longer perceive it. That is the basis of why we are no longer in touch with the consequences of our actions and are able to ignore them.

Arregoces Coronado-Zarabata shares:

Why isn't it possible to bring the Zhatukwa to Europe? If the Zhatukwa is there and the other Mamos are here, that's not good. Every area comprises a spiritual boundary. So when I am there and ask about the rain, the Zhatukwa will always point to the rain that is here in the Sierra and not to the thought of the rain there in Europe. Why? Because the Zhatukwa is registered here and was made to function here in this area of the Sierra. It can't just be registered somewhere else. Sometimes when you see a sacred place, you do not know what this sacred place corresponds to. Whether it is a place for the water or for the wind or if it is the place of the trees. When I was traveling with Mama Shibulata, he simply looked at the places and knew exactly what that specific place was for and who had to work there.

Even if this analogy with the Oracle is a general one, the Zhatukwa, as a distinct form, is rooted in a specific area. The same applies to food, houses, and objects of all kinds. For us, this is very difficult to imagine, since we are so detached from places. Unlike us, the Kogi do not see thoughts and ideas as existing exclusively within the realm of the human psyche, but as forms that are also independently existing in space.

We Younger Brothers and Sisters live differently; the Kogi emphasize this again and again. For us, technology plays a greater role, and our lives are less bound to just one place. Therefore, it is even more important that we maintain our connection to the places in our life. When the person is not present any longer in the place where they took something, that place is then unable to receive their balancing energy and can become sick. Since everything is interconnected, the illness of one place also inevitably means the illness of other places and therefore also the illness of people.

THE KOGI TERRITORY HAS ITS OWN LIVING INTERNAL STRUCTURE that consists of places and zones with different objects, tasks, and worlds. In its entirety, it is encompassed by the Linea Negra, which, much like a spider web, consists of traversing lines that run through the land and also give a clear definition to the border of the territory.

Mama Bernardo Mascote Zarabata explains:

The traditional territory is composed of three essential parts: the *esuamas*, the sacred places, and the Linea Negra. It is governed by an ancient traditional order and requires constant guarding by its inhabitants. To accomplish this, they have to act in a certain way and take on their responsibilities. This is the strength and essence of us Kogi, of us humans. For this reason, before we do anything, we order spiritually and physically the three component parts of our territory: *esuama* [place of power and knowledge], *kaguí agzain* [sacred place], and Senénulang [the territory]. In the lower parts of the mountain range, on a physical level, it is the mountains, lakes, rivers, and streams that limit our territory. At the center of the range is the mountain Gonawindúa, whose name has a very deep meaning.

The Younger Brothers even go to the lakes and work there with large machines. Do they really believe that the Mother, the earth, has not organized it well? Or why else do they think they have to change something? If they don't think that, then one explanation is that they enjoy destruction. Why do people who are not from here come to walk on our mountains? They have their own mountains, even much bigger ones than ours. Why don't they walk around there? If they walk here, they are acting against our laws. It's like a stranger just walking into someone else's home. He should not have been there. We do not want any more Younger Brothers wandering around here without our permission!

WALKING MEDITATION

The Kogi have specific practices to maintain their connection to the land. The Kogi Mamos continuously study the structure and condition of the territory in all its details. They do this by sensing and feeling as they walk through the territory. Where are the sacred places? Which places are meant for living and dwelling? Which places are for working and which are the fields? It is about both the physical land and the thoughts that are located there. The ordering of a sacred place is also always the ordering of the community and the ordering of the individual. The ordering practice in one place has an impact on the whole territory, yet it is important at which specific place the ordering of the whole is done. I have experienced Mamos who explained that they now had to walk to a place two days away for some matter because the task simply could not be solved here on the spot. I even observed Mamos going around the village, asking if anyone else had something that needed to be brought to that specific sacred place, so that the walk would be even more worthwhile and they wouldn't have to walk back again so soon with another situation.

Some of these places are culturally significant, such as the places where the *poporro* is given to the adolescent, places for the ceremony of the newborns, weddings, and death ceremonies. It depends on each individual, because just as these activities can strengthen the territory, they can also be the reason for disorder. The smallest actions in people's daily lives are of the utmost importance and serve as the foundation for the functioning of Senénulang.

The relationship between human beings and nature is very

much a reciprocal one in the eyes of the Kogi. For us to obtain a similar understanding, it is not a matter of recreating their system and way of doing things, but of paying greater attention to the subtle and delicate connections that exist between us and all living things.

"SE DICE *KAGUÍ AGZAIN, KAGUÍ AGZAIN*, SITIO SAGRADO," SAID Mama José Gabriel, and I repeated after him. When pronounced, it sounds like *kaguí arsén*. The syllables are sticky and taste almost like honey on my tongue. So I repeated the words *kaguí agzain*—sacred place—two or three times, until the Mamo nodded contentedly. Sometimes the sacred places are also called *nujwákala*, which means "cave."

It took me a long time to understand exactly what sacred places are for the Kogi and why they are so significant for them, and I soon realized that my understanding and image of a sacred place simply did not fit. No matter which Mamo I asked, I was always told that in the sacred places Sezhankua is present and will fertilize these places to make life on earth possible. Then it is the responsibility of the Kogi to maintain the continuation of life in these sacred places. Also, sacred places always hold sacred objects that keep the earth and her beings alive.

These words made sense to me, but at the same time there was a depth to them that I could not fully comprehend. I had to ask myself the questions:

What does this have to do with us?
Is this simply a cultural view of the Kogi that bears
little relevance for us?
Or can we learn something from it that will help us?

My first, simple conclusion to these questions was that their completely different view of land, earth, and territory alone has the ability to enrich our modern, urban view.

Sacred places exist as a network. Alone, none of the sacred places in the Sierra can fulfill their function of ensuring the continuation of life and of the various living beings in the heart of the world, and thus in the world as a whole. According to the Kogi, the inherent qualities of a sacred place are only accessible when the place is energetically connected and in communication with other sacred places.

The communication between the sacred places in the upper mountains and those of the lower mountains is important for the maintenance of the entire territory. If this contact is interrupted, the traditional balancing work is no longer effective. It becomes more difficult for the Mamos to integrate both the spiritual and physical dimensions so that no discord or separation occur. The blocked energy creates difficulties for both the individual and the society. Therefore, the Kogi say that having functioning sacred places is the basis for the order of every territory.

Whenever the Kogi talk about people, they always talk about the land, and that works both ways, because for them the two are inseparable. If we don't appreciate each other, we fail to appreciate the land and vice versa. Just as the inability to acknowledge the sacred in each other mirrors the inability to recognize the sacred places. If we take from the land without reciprocating, then we also do the same in our interpersonal relationships. In this respect, one can understand why the Kogi correlate conflicts and accidents to the relationship with the land and places. In other words, when I disregard an aspect in my life, I also disregard a corresponding place within myself.

The Kogi know two types of sacred places: general and specific. The former are mainly located along the paths through the river valleys and in the mountain ranges. These places are used by all Kogi and serve many functions in life. Sacred places can, for example, be designated individually as places for all birds, insects, or the general balancing of energies. Some specific sacred places with clearer and specific tasks are only accessible and guarded by the Mamos. The Mamos work at these sacred places with specific types of animals or plants or with certain principles and energies. The Kogi say that in this way, they balance natural phenomena like heavy rain, landslides, and earthquakes.

THE ACT OF BALANCING

Mama Ramon Gil Barros speaks:

Everything has its own place. If many chickens or many monkeys die, the Mamo goes to the sacred place that corresponds to the chickens or monkeys and balances the disharmony. Then fewer will die. When the turkeys are decimated, he heals that corresponding sacred place. If there are many snakes, we go to the place and do our work in order to restore the balance and to find an agreement with the snakes so that they don't come anymore. Then they truly disappear. In case of a heavy storm, the Mamos visit Teyuna, which you call the Lost City, and visit the stone of the frogs; it is called Kuizbankuish. There an agreement with the situation is found, balance is restored, and therefore healing takes place.

The sacred places of each and every thing are equal. The people of the Sierra make their way up the mountains and carry objects of physical and spiritual nature, called *zabiji*, which are used to balance the disorder we cause by living in nature. For example, from the beach they bring a levy, called *shémake*, as an offering for the girls and boys. Farther up in the Sierra, they collect *msima*, which is used to provide balance for the girls. Even farther up there is the *zhukatá* for the *poporro*, and even farther up in the mountains there is *kwalama*, which brings balance for the children and the seeds. This balance takes place energetically, spiritually, and physically.

Every situation is based on an energetic resonance, which, for the Kogi, corresponds to a specific place, similar to reflexology, where specific zones correspond to different organs. What comes from a specific situation can be eased, or even dissolved, by acknowledging and accepting the energetic resonance of that certain situation. To do this, the Mamos visit the sacred places where the origins of the respective energies are located. The reason for going to a sacred place can be anything from dealing with envy to hoping to cure a cough to the desire to have a child.

The Kogi say that they speak to the mothers and fathers of things that are the origin of an individual situation. Therefore, when the Mamos go to a sacred place to make a *pagamento*, it is their way of acknowledging and accepting the situation to create balance, much in the way that one listens to a loved one and accepts their view as real, so they can create balance between views.

We might ask ourselves what the word *sacred* actually means. The idea of the sacred has almost completely disappeared from the lives of many modern people. For us, the term *sacred* has religious connotations and is often used in association with the church.

For the Kogi, however, *sacred* simply means "untouchable," without any sacral-ceremonial connotation. Untouchable because nobody is allowed to enter these places except for a few responsible Mamos or Sakas. Only they are familiar with the essence of the place to the extent that they can handle the energies and thoughts that are accessible there responsibly.

WE CANNOT CONNECT WITH OTHERS IF WE DO NOT CONNECT WITH THE EARTH

To the Kogi, places are just as alive as people, and therefore they can communicate and converse with them. The Kogi even go as far as to say that we cannot establish connection with another person, for example, our partner, if we do not manage to do so with the earth.

Mama Bernardo Mascote Zarabata explains:

There are many sacred objects that the Mother herself placed in an area where no one can reach. One cannot go there. However, some people are curious and still want to go to these places and touch and see these things. No one should walk on the snow-covered mountains here in the Sierra. If one walks in these places, one's thoughts can damage these places, because these places act like amplifiers for thoughts. This can lead to stronger imbalances. The Mamos who have been practicing confessions, *alúnayiwási*, for a long time are the only ones who are allowed to visit these places. These places are only there to be guarded and protected.

Today we are talking about the most urgent issue of our time. It is not good to take the gold out of the earth or to pierce the earth with mines. The Mother has left us the territory in a certain way, with the request that we must not destroy the earth. Sometimes people seem to be eager to destroy the earth at all costs by extracting mineral resources and by digging up archaeological graves. We do not understand this. We live in the Paramo, beyond the timberline, but we would never attempt to settle and live on the shores of the mountain lakes. We only visit these places very briefly, in order to do our spiritual work, and then we return back to our homes. We never throw stones into the water. We would never build a house on the shore of a lake. Why? The Mother has decided it to be like that and we listened and understood everything. Our Younger Brothers want to take and own everything they find. We have made it clear so many times that the estuaries of every river are sacred and that they must not be altered or destroyed under any circumstances.

These lakes are the strength of the Mother. The Mother breathes through them, they are her lungs. I have now seen this lake dry up. This lake is connected to another lake, which can be found high up in the Paramo. These lakes are the mothers of the rain, they call the rain. When this lake still existed, the cycle of rain and drought had not yet been affected; they were in balance and it rained when it was supposed to rain. Until then all followed its original order. But nowadays, the Mother is like a sick person, she only breathes with one lung and is very weak.

Are you surprised by the fact that the Mother also dries out when you dry up the lakes? The Mother becomes thirsty. All of these sacred places are connected. When we make a pagamento in one place, it has a rippling effect. We still do our work in these places, but the special Mother has disappeared; nobody is present there anymore. In San Miguel, a young man told me that some Americans had climbed a mountain where not even every Mamo is allowed to go. From there they took some sacred stones that are a source of strength for the whole universe. These stones give strength to the mountains, but now they are no longer there. They just took them away. Nobody without explicit permission is allowed to climb that mountain, but they simply did it without any care. After that, when the Mamo went up the mountain to make a pagamento, the fathers and mothers had disappeared and no one was there anymore to receive it.

We used to be able to make good pagamentos for a woman's pregnancy. In the past, Kogi women never had problems with pregnancy, but nowadays, it is getting more complicated. Why is that? You have been mining a mountain that is a woman. Her intestines have been damaged and her organs can no longer function well. Today, even children are sometimes born sick. How can they be healthy if Mother Earth herself is sick?

The Mamo tells us that we, the Coronado Shimangue family, are the family that has been assigned to be in contact with the Younger Brothers; we have to communicate with them. We have already been talking about preserving the earth for some time now, yet it seems that they simply do not understand. Everything was primeval forest, but they themselves cut down the trees. The trees can recover. You can leave the forest alone for a while and it will start growing again. For us, the hollowing of the earth is much worse. The drying out of the lakes is also fatal. These things are lost forever. This is the first priority when we speak about the protection of nature.

Many people think that only we natives suffer from these things, but that's not true; you will also get these problems in your countries. There might be no drought but big floods. And there will be diseases and violence on your continent if you don't change anything. This will not be far in the future, it will happen soon. In parts it is already happening. Things start here in the heart of the world and spread from here. Because it is like in the human body, the blood flows from the heart to all other parts of the body. The strength and vitality of the Mother is but one energy. It is all-encompassing and permeates everything. You Younger Brothers often do not understand how the world is constructed because you only pay attention to what is visible. When you see a river, you think that it only has importance for the place where it flows, but it is connected to everything. A change in the river causes changes in a completely different place on the other side of the earth. Everything feeds off each other. Because so much has been changed and damaged, the Mother's strength is no longer the same.

The Kogi's most important message to the Younger Brothers and Sisters is to leave all sacred places alone. Tirelessly they kept repeating this point, as well as the problem of extracting mineral resources. The issue of mining mineral resources is not necessarily easy to solve, yet it is at least easy to understand.

As for the topic of the sacred places, it is a little more complicated, more difficult to understand, since we have lost our ability to recognize and establish a connection with them. Places act as amplifiers for thoughts. Therefore, we may want to ask ourselves which are the places where we spend a lot of our time?

Which places do we choose for work or to meet friends?

What do we love about these places?

How do they influence us, and what does our presence do to them?

Mama Ramon Gil Barros offers:

Many sacred places along the Sé Shizha [Linea Negra] are barriers that keep natural phenomena, hurricanes and disease, in the places they belong.[3] That is why the sacred places play a fundamental role in the prevention and protection of health. Some Mamos work to protect different materials that we use for our traditional work. We collect medicinal plants in these places, which don't grow in any other climate zones. Many plants are similar to plants from the high mountains, and those we use in the sacred places of the higher planes.

There are special places where we take care of the spiritual imbalances that the Younger Brother has caused, because our work is for all people on this earth. However, in spite of our efforts, the destruction continues daily in the form of large-scale projects that complicate the situation more and more. We do not understand what Younger Brother wants to achieve with his actions; he doesn't listen, he doesn't want to understand. It will never be possible to apply a positive intervention in our territory without the different elements and energies being in balance. To restore balance requires a very deep study of the region where the work is to be done. It is simply not enough to eliminate the reasons for destruction and its consequences to restore whole habitats.

Instead, it is absolutely necessary to know the function of

each individual place, so that the respective tasks they play in the system can be understood.

When the Kogi talk about places, they mainly talk about places that exist within their territory. One of the largest coal deposits in the world, El Cerejon, is situated at the foot of the Sierra. The coal is extracted from an enormous mine, releasing toxic chemicals into the environment and destroying vast stretches of land that now resemble a lunar wasteland. At the same time, huge loading ports were built on the coast. Of course, this has far-reaching consequences for nature. We need to understand this to comprehend the severity of the Kogi's message and understand that what they speak of has a direct consequence for nature, and therefore for all of us.

CHAPTER SEVEN

What Does It Mean to the World That I Am Here?

The doorway of the *nuhué*, the ceremonial hut of the Kogi men, is low and I must bend down to enter. Inside it is dark and I can barely see anything, yet slowly my eyes adjust to the darkness, and the faces of those present begin to emerge, shimmering in the faint glow of the dimly lit fire. I can vaguely guess who each person is. This large wooden hut, as ancient as it is simple, represents the structure of the universe and is home each evening to the conversations of the men, and also the place where they sleep. On the other side of the hut, I see an empty hammock, so I go and take my space. It was clearly made for someone much smaller than me, yet I soon find a position that allows me to settle in and immerse myself in what is happening around me. The warmth of the fire, coupled with the darkness, makes me feel safe and deeply peaceful. It is always dark in the nuhué, and the fire glows. The darkness feels at once like being held within a womb and yet infinite and vast.

Every evening after dark, and after having had dinner with their families, men of all ages gather in the nuhué, while of-tentimes the women will gather in the *huitema*, the female ceremonial hut. In the nuhué, heavy clay pots are placed on each of the four fireplaces, and from one of their bags, the men take the coca leaves that their wives have harvested for them during the day and begin to dry roast them.[1] This is an art that a Kogi man will master over the course of his life. The leaves must be neither too moist nor too dry and especially not roasted too hot, but in such a way that their delicate green color remains. Only the men chew the leaves, because accord-ing to the legend, the coca plant is the Great Mother herself. When men chew the leaves, they connect with the feminine aspect and therefore balance excessive masculinity that could otherwise express itself in destructive aggression. The Kogi explain that women do not need to chew coca leaves because they do not encounter the same risk of imbalance.

As the evening turns to night, the nuhué is filled with an ever-increasing ethereal haze of smoke and sound. I notice that the men have finished roasting their coca leaves, ready for use the following day, and have begun to work with their *poporros*.

Time gently evaporates as the conversations around the fire continue well into the night. The talk is often calm, but it sometimes veers off into an agitated tone as the passion of the story that is being revealed requires those present to listen. While stories and myths are shared, the day, and what took place, is also reviewed. Once there is an agreement, the next phase is to consider the following day and what that shall bring. The Kogi talk in hushed and mellow voices, and during the night sometimes they gather in small groups of two or three men, while at other times they are all together in a large, harmonized, group. Sometimes one person speaks at a time; in other instances a few people speak at the same time and separate conversations unfold.

Here, in the mystical space of the nuhué, the questions echo in the dark, and it is often impossible to tell who is talking. Though regardless of who may be speaking, it is always all about the stories, which sometimes can go on for hours. The stories are often about the ancestors, the Tayrona, and about the dawn of time. The Kogi also talk about their daily lives, their tasks, and what they will do tomorrow. There are moments when nobody talks, and this allows for what has just been shared to sink in. In the darkness of the nuhué listening becomes thinking, because for the Kogi both are the same. Whether they are listening to each other's words or to the sounds of nature and the fire, it is always stories that they hear being told. Whether they are listening to the sound of each other's words, or to the sounds of nature, or the fire, or rain, the Kogi hear each sound as a story that is being told. Each story is a thought that is brought from the invisible realm into the physical world with the meditative motion of the wooden stick of the *poporro*.

I rest in my hammock during a break between stories and attempt to absorb my surroundings, and the otherworldly scenes I am witnessing. There is a dance of energy in the air, something that is impossible to put language to, a quality that pulls me into the present and consumes any ideas I might have of truly comprehending where I am. As I lean back in a gentle appreciation, one of the elder Mamos walks toward me and sits beside my hammock. He looks ahead into the heart of the fire and begins to speak to me. Kággaba, the language of the Kogi, fascinates me. The language has weight to it, and is sweet and raspy, yet somehow soothing.

We ask ourselves important questions, questions that bring us together, so we may think and sense as one, "Are you sitting well? What did you plant, what did you harvest? Do you know the stories well? How will we pass the knowledge on to our children? What do I have to offer the world? What does it mean for the world that I am here? What thoughts have you been thinking? How will we make sure that everyone is well? How will we live another eighty thousand years?"

A silence drifts between us for some time as I let the questions float around me, and I watch the depth of the orange embers of the fire breathe in different tones as they transform from each gentle gust of air in the nuhué. After a while, the elder Mamo, still focused on the fire in front of him, says one last thing: "We ask ourselves these questions so we may change our thoughts, so our thoughts may change the world." Soon after, he rises and walks beyond the glow of the fire, disappearing into the darkness of the nuhué, leaving me to contemplate what he had just shared.

This task is what the Kogi see as guarding the earth, and this understanding of themselves offers all humans an immediate value. How would it be if we would also naturally, beyond any political or religious moral doctrine, but in the spirit of holistic balancing, be responsible for the thoughts we nourish and give attention to?

ONCE AGAIN I AM SITTING IN A HAMMOCK IN THE NUHUÉ long after dusk; the Kogi men have gathered around the four fires, their faces touched by the soft glow of light of the orange embers. They talk about their day and which thoughts influenced their actions and about what tomorrow might bring. Their calm voices, merged with the cadence of the *poporros* and the radiance of the crickets outside, soothe me into a dreamlike state. I feel as if I have once again stepped across a threshold into another world, one of raw myth and sensation. In the darkness of the ceremonial house, I begin to find that everything seems more heightened and my feelings more intense, and gradually, despite my best efforts, I can no longer keep myself awake. I had already dozed off once or twice, and finally I decided to just surrender to the dark and curl up comfortably in my hammock and fall asleep. In the background I can still hear the murmur of the Kogi's gentle conversations flowing for a long while.

When I awake the following morning at just before six, the nuhué is almost completely empty and the men have already returned to their houses, where the women will have likely been preparing breakfast since before dawn. The Kogi men and women sleep very little, on average only three to four hours a night. For a long time I could not understand how this was possible considering how energetic and strong they are. Over the course of my stay, however, I realized that this is supported by their relationship to darkness.

The Kogi say that ideas and creativity emerge best in the deep black darkness. In contrast to our world, darkness is something positive for the Kogi, and I even get the impression that they feel significantly more comfortable when it is dark. In the nuhué, they enter the darkness and the world of the unseen in order to give meaning to the visible. In modern society we often forget that while the visible and the invisible worlds are different, they are interconnected and inseparable. In Kogi tradition the world of darkness is where everything is written, where all laws of life have their roots, and where things actually exist. For the Kogi, what exists in the visible world is merely a reflection of what has long existed in Aluna, the world of thoughts. The world of light is the world in which we often forget what is essential and are blinded by the phenomena and appearance of things.

There is an intrinsic coalescence between thoughts and darkness, and they hold the potential of all possibilities. Jaba Sé and Jaba Aluna, the mother of darkness and the mother of thoughts, are the original creators of the world.

Mama Bernardo Mascote Zarabata speaks:

In the origin, the Mother organized everything very well. To exist at all, we were placed in the water. This was the decision of Jaba Sé, and it was accomplished in different stages. Jaba Sé existed long before the sun was in the sky. At that moment, everything was in darkness. That was Jaba Sé. At the same time also Jaba Aluna, the mother of thoughts, existed. Neither was there before the other. For us, Sé is the time when we are in darkness and still in the water. A child in its mother's womb is still in the Sé, but it also has Aluna, it has thoughts.

Arregocés Conchacala continues:

We natives always order our thoughts before we act. This is the reason why someone goes through a specific thought process before traveling to another village, before building a house or planting something. Before doing something in a material way, it is important to do the inner organizing. Once we achieve this, then we have the ancestral permission; this is what we call sewá.

Mama Pedro Juan Noevita adds:

There is nothing but the spiritual worlds that materialize themselves into matter, because everything we need already exists in thoughts. Jate Shilkuankan began to build a house for Jaba Sé. He told her she could live there, but it was badly built and it collapsed, and he had to build it again. That is why Jaba Sé created a cosmic law, which determines how to create a house and how to organize oneself so that it materializes well and one can live there. Jaba Sé has built the house for us to order our thoughts, to organize and make decisions. This house is the nuhué.

When talking about Aluna as a world of thoughts, it is important to understand that it is not simply a concept, but really is a world that exists in parallel to ours.

AFTER ONLY A FEW DAYS OF LIVING IN THE VILLAGE WITH the Kogi I was able to fulfill my most basic needs, and I had begun to get used to life there. I felt that the Mamos had also begun to accept my presence, even though I could still sense an air of distrust over why exactly I was there. Yet, often they would come and speak to me. One man asked me if I had come to buy their land and build a mine. I explained to him that I was there to listen to the Mamos and Sakas and to hear the ancient stories of his people. He looked at me a little suspiciously, was silent for a moment, and then said "*wuà*" as an expression of approval.

This simple interaction sparked a long conversation between us. The man was very interested to hear of the ways in which we live in Europe. I showed him a few photographs from Frankfurt, where I lived, and the landscapes in Germany.

He was amazed. "You also have nature in Europe!"

"Yes, of course, why are you so surprised?" I asked him.

He then explained the following to me: "The Mamos say there is nature all over the world, but I somehow couldn't imagine it. You Younger Brothers only think about money and machines, but not about the thoughts of the Great Mother, that is why I assumed you have no nature there."

His words touched me. For the Kogi, thoughts are not merely abstract information, but living beings, children with mothers and fathers that procreate as living beings in this reality. Thoughts move through space, request specific things, and can, like a lost child in a shopping mall, be brought back to their parents if they become too unruly or get lost in places where they don't belong.

THOUGHTS MOVE INDEPENDENTLY

Mama José Gabriel explains:

Thoughts are like stray dogs. If you meet one on your way, it starts to follow you and you can't get rid of it.

Somebody may ask: I don't see them, how can it be that the

thoughts are there? These thoughts enter the hut at night. They wait in the doorway for someone to take care of them. Then, when you sleep, they come to you.

That is why the Mamos are awake at night and take care of their own thoughts. If we think well and speak well, then good thoughts come to us, but if we think badly and speak badly, bad thoughts come to us. Whatever feels at home with you will come to you.

When someone cultivates their field, builds a house, has a family, then these thoughts are carried on and the children receive them too. Negativity only really enters our thoughts when we are asleep and don't realize it, while positivity can also come while we are awake. Your own thinking determines which energies have access to you. If you think badly, you open the gate for negative things, but if you think well, you open the gate for positive things that come to support you.

The reason for someone's sickness is always linked to their thoughts or actions. However, when they go to the state-run health center, they give them a pill and promise that it will make them well again, but the negative thoughts remain and the problem is not solved. This is not in accordance with the Ley de Sé, and therefore after a short time they simply get sick again because the cause has not been eliminated. We don't need pills to get well, we just have to clear our thoughts. This is also a form of *zhigoneshi*. Thoughts naturally multiply, the good as well as the bad thoughts.

Some of these principles might sound familiar to us: the fact that negative thinking does not have an optimal effect on our health, or that thoughts "occur" and that we do not really "make" them happen, and that it makes sense for us to focus on what is positive and supportive in our lives.

The way in which the Kogi truly live and apply this wisdom and awareness in their daily lives has time and again amazed me. Their views are not simply a philosophy that is shared for entertainment or to fill time during long evenings spent by the fire; their views are the fundamental basis of their culture. They absolutely cannot understand why we do not consider questions more often in our daily lives. Yet, their advice is not only to ask all these questions to bring balance and clarity to our lives, but they also advise us to return thoughts that are not our own and give them back to where they have come from. We have strayed far in our modern culture from the origin of things, but it is now more crucial than ever that we once again acknowledge the origin and the principles of life.

We should be asking ourselves specifically:

Whose thoughts am I thinking?
Where do they come from?
From my parents?
From a friend? From my boss?
From the media, society, or the church?
Or are they truly my own thoughts that belong to me?

CARING FOR CHILDREN FROM THE BEGINNING

The Kogi apply the cleansing of thoughts to all areas of their life, and this process begins long before a person is born. Children play a very big role for the Kogi as they represent not only new people but also new worlds. Therefore, the care and order of this new world already begins during pregnancy, because children are strengthened or weakened by the thoughts of their mother, as well as by those of other family members. Good, healthy, and strong children are a sign of good thoughts. Thinking, nourishing, and fertilizing are all the same for the Kogi. Good thoughts are actively cultivated on a constant basis and are not just briefly considered, because they are living beings. For the expectant parents, awareness of thoughts is an essential component for the successful growth of a child, which includes the process of ascertaining the child's unique task in life.

Mama Bernardo Simungama Mamatacan speaks:

It is important to begin the spiritual work with the child when they are still in the womb. Through the birth of a child, a new world is born, and this world must be created well. It is like a house: if it has a poor foundation, it will quickly break down and cause damage later. We would not eat fruit or food of bad quality, and it is the same for the child.

If we already think well for the child in the womb and consider its individual task and that it will accomplish it well, then we keep the child free of negativity. Everything we do is energy.

The children in the womb feel everything and absorb everything. For us, the children are very sacred because every child is a new world. They are the fruits of women, just like the trees bear fruit. We treat the children very well and think well about them, because this is the only way we can be sure that they will be healthy and strong. When a child is born, it is a new life that fulfills a certain task for the Great Mother.

We have heard that some of you believe that one can be born more than once. You call that reincarnation. We don't see it like that. All children who come to us are new, just as every fruit that a tree produces is new, but it always comes from the same tree. These fruits have never been here before, but the thoughts that the child brings are not new. These thoughts already existed before. When a child is born, the Great Mother is rejuvenated. The birth of a child also reinvigorates the ancestral lineage, such as the one of Sezhankua or Siukukwi.

We always talk about the number nine. We have to do everything for nine days: nine days of initiation, nine days of purification, nine days of ceremony when we receive the *poporro*. The word for nine is *itagua*. For us, the number seven is a negative number. Negative animals and negative thoughts have the number seven. Four is another important number for us. The world has four pillars on which it stands. Each chair has four legs, there are four fires in the nuhué, and there are four directions. We Kogi also live with three other peoples here in the Sierra Nevada, so we are four peoples maintaining the stability here. That is why the fewest number of days we perform a ceremony for is at least four days and four nights.

Children are the fruits, they are the consolidation of everything we maintain. Life springs from the plants. We eat bananas
and sweet potatoes, but also *ñekes* and *guatinajas*. Therefore,
we do our work for the plants too, for nine days and nine nights.
We are not different from nature, we are nature. There is nothing
else. Nothing exists outside of nature. We eat what comes from
nature and then become nature. Therefore, the children who are
born are the sisters and brothers of the birds and plants. They are
also relatives of the wind and the water, who are our grand-
fathers, just like the stones. This has always been so and only
functions as a unit. That is why a pregnant woman has to come
regularly to reveal her thoughts to a Mamo in order to protect the
child from whatever could affect this unity.

In the language of the Kogi, the term *human* has a special
meaning: the word *kággaba* means "human" and the corresponding verb *kagbei* means both "thinking" and "creating."

In reference to English, for the Kogi the two words
thinking and *creating* are indistinguishable. It is also important to note that the Kogi do not differentiate between sensing and thinking, and that the latter does not imply a purely
mental process but a conscious perception with a form that
is alive.

Therefore, as a being that thinks, humans have access to
Aluna and also to an enormous amount of creative possibil-
ities, although these must be realized in accordance with the
Ley de Sé. For instance, the Kogi say that it is absolutely

accurate for the Younger Brothers and Sisters to construct and materialize technologies and machines that they have conceived of in the realm of thoughts. However, these should be created in accordance with nature and the cosmos, that is, in the One Thought, and not as it is currently: convoluted and destructive. In their work with, and in, Aluna, it is essential for the Kogi to thoroughly allow the intricate process its time in order to have a full understanding and overview of any potential consequences. This is also a place where they see the Younger Brothers' and Sisters' lack of care and attention.

The Kogi see that our difficulties do not only arise from taking impulsive action without thought, but because we do not adhere to the important and vital thoughts we do have. To assume that somebody else will solve the problem for you is to neglect one's own responsibility and results in becoming absent and apathetic about one's own place in the structure of life. Many traditional stories of the Kogi speak about the fathers and mothers of all things realizing that the effective continuation of all life depends on the materialization of thoughts.

On the eve of an important meeting, the Mamos sit together, and in Aluna they conceive the exact process of what is to come. This process is about perceiving the impulses of the world of thought rather than an arbitrary visualization of their wishes and ideas. During this process they work intensely with their *poporro*, which connects them to the principle of implementation and manifestation in this physical world. The Kogi say that the movements of their hands with the *poporro* are movements in Aluna. In order to access a meditative state of manifestation Kogi men work with the *poporro* several times a day. It serves as a constant reminder of their role as a human being: an active co-creator of the world. Being a creative participant in the events of the world is an essential part in expressing their innate nature.

Mama José Gabriel speaks:

In the beginning, everything was only thought, but Sezhankua and Jaba Seinekun said that they will change this. Before that there was nothing but thoughts, then we became water and finally became flesh and blood. When a child is born, it cries but it does not speak. As it grows, it starts to talk, to work, to build houses.

The thoughts, however, were there long before. Sezhankua saw that the world was not good without people, so he created us, but we are all children of Sezhankua. Everything existed first in Aluna. Even today we still think about what we will do tomorrow, what we will eat tomorrow, and you think about where you will get tomorrow's money from.

Thoughts never change, they always remain the same. First, we think about going to a certain place, and only after do we actually go. Or we think that in the afternoon we will take a bath, and only after that do we do it. The thoughts have existed since the beginning, this is how Sezhankua entrusted it to us.

There was a time when we only fed ourselves on thoughts. Then we planted food and began to eat it, but before that we had already thought of the idea of eating the food; why else would we have planted the food? Now we eat everything that grows, but the original thought to grow food came from Sezhankua. Just as we were first made of water and now of flesh and blood, in this way everything exists first in thought and only after in matter. A man who thinks alone doesn't get anything done, he must think together with the woman, then something can be created into being.

When a woman is pregnant, she carries the child inside of her, and exactly in this same way she is carrying the thoughts inside and lets them grow. That's why men live together with women. The children then come to both of them, it is only in this way that their thoughts will live on. We cannot allow for these thoughts to be forgotten; they must be passed on.

For the Kogi, creativity is the manifestation of already existing possibilities and is not merely a reshuffling of the mind. When the Mamos enter Aluna, they connect with the invisible world of possibilities, where all problems, as well as their solutions, already exist. There cannot be a problem without a solution because every imbalance already contains the components for its equilibrium. In this case, it is necessary to recognize and apply an already existing simple solution instead of overanalyzing and deeply examining a situation that often separates the problem from its solution.

Juan Mamatacan explains:

When the *poporro* is given to a man, the ceremony takes nine days. During these days the young man concentrates very well. This concentration is the same concentration that he will need to take good care of his future wife and family. If he drops the *poporro* and does not concentrate well, then this is a bad sign. In these nine days, without sleeping, the young man conceives in thought his future life as an adult man. He thinks about every-thing. He also thinks about what he will refrain from, that he won't steal, that he won't take another man's wife, that he won't fight. During these days, he is not allowed to walk around or speak to anyone other than those who come to him to give ad-vice and to encourage his thoughts. He sits quietly. He listens to the Mamos.

The girls sit for seven days when they become a woman. The girls, like the young men, are then brought to a cave or a large nuhué. There they sit and think and think and think. The elder women then come to them and give advice and counsel.

The word *concentration*, for the Kogi, means, first and foremost, the absence of doubt and distraction. This allows for an open presence and connection with a subject or an energetic quality. For them it does not represent a goal-oriented tunnel vision as it often does for us. The time they dedicate to their rituals gives them the possibility of gain-ing real clarity of thought, which allows them to consider the consequences and integrity of their actions as well as how well they are implementing them. They are focused, perceiving, and connected, not separated in intense concentration.

WE CAN'T JUST WANT, WE MUST INVITE

Mama José Gabriel shares:

We work to call the rain. We were taught by Mama Kakamunkue at the beginning of time. We search for the clouds and invite them to come to us, and then it rains. We invite them in our thoughts, and then they like to come. When we simply want them to come, nothing happens. The old Mamos say that all thoughts have existed since the beginning of time, but one must think well and thoroughly. If one's thinking is scattered and fragmented, nothing happens at all, that's why some people assume it doesn't work. All aspects of life must be considered: the trees, the bushes, the frogs, the snakes, the birds. In our thoughts we consider that all these beings also need water so that they don't suffer any more either. We take care of them, then Sezhankua also helps us. If we are aware that we are all brothers and sisters of the trees and the animals, then it will rain, but if we just call the rain only for us, then again nothing will happen. Also, we always first ask for agreement with the situation. If I think well, then I will be heard on the other side, but if I argue or speak badly or even criticize anyone, then I am not on the path of Sezhankua. We have to guard everything very carefully.

JUAN MAMATACAN COMES FROM A VILLAGE IN THE LOWER regions of the Sierra. In his village, a school was set up by the Colombian government where the children learn to read and write in Spanish and are introduced to Colombian culture. Even though the school does employ some Kogi as teachers to teach their own language, in both written and spoken form, the Kogi have mixed feelings about the school as they observe how their children are changing. Some of these government schools have already been closed as the Kogi no longer want to send their children there. One of the dangers of the school that the Mamos see is the mixing of ideas from different worlds, which will result in the disappearance of their culture.

Juan Mamatacan speaks:

We natives must not mix. If we do, we will lose our culture and above all our thoughts. Thoughts must complement each other and not mix. The children here started playing war because they had watched such films in their school, and now they imitate it. The responsibility for the children's thoughts rests with the fathers and mothers. The children will listen to the birds again and hear what they have to say, and will not watch films in the school of the Younger Brother anymore. Sometimes the birds are cheerful, sometimes they call the rain, or they speak of something. It is not only the Mamos' task to teach the children, every father and every mother must do the same. The Mamos help to gather and process the negative thoughts of the children and to stimulate their good thoughts.

We explain to the children that there is no such thing as a game. What do I mean? In a game, if, for fun, I say I'm going to kill someone, for us, I've really killed them, because we feel it. Imagine how many people have killed someone, just by what they have thought. We all have to bring these thoughts to the Mamos, otherwise there is no way to bring back to life those who have been killed in our thoughts. The mothers and fathers of all things assist with this. The children who live very high up in the mountains are still different, they have never seen a television in their lives, and they have never seen a Younger Brother. When they meet a stranger, they don't come and speak to them like they do in my village, but rather they hide. They also play differently; they imitate the adults; they build small houses with sticks. The artificial thoughts here in our children will multiply if they continue to go to the school of the Younger Brothers. The children up in the mountains are very dark, because their clothes are not washed clean. The Mamos of the high Sierra see the children down here in my village as having too little of Mother Earth on their clothes.

IT WAS A HOT AND SUNNY AFTERNOON IN THE SIERRA, AND we were sitting in our hut with Mama José Gabriel, when an unexpected opportunity arose for him to teach us one of the most important principles for the Kogi—the value of acceptance and the balancing of ones thoughts. We were immersed in conversation, when suddenly from nowhere a Colombian man entered the hut and interrupted us: "I would like to introduce myself. I'm an engineer and I've been sent by the government to take a look at this area." He had brought a box of doughnuts and offered them to us. Mama José Gabriel looked at him for a moment, then smiled, thanked him, and cheerfully took a doughnut. I was rather hesitant at first, because here in the Sierra the sickly sight of a doughnut felt very alien to me, and I was also concerned that the man was here to look for opportunities to build a mine and was using the doughnuts as a cheap bribe.

Once the man had left, while enjoying bites of his doughnut, Mama José Gabriel explained:

Here you can see how life functions. We speak the original words of Sezhankua. Who told the man to bring us food? Sezhankua himself did. He knows that we are hungry and he sent us food. We do our work very well, and this is a sign. If you are in a conversation and someone comes to bring you a meal, you must never reject it, even if you don't like it; this is how the traditional story has been told to us. Because if you do that, you criticize life itself. It doesn't happen often that someone comes here to bring us food. We are doing very good work. Sezhankua himself sent the food and therefore we have to eat it. We receive compensation for the work we do and what we create, that is the way it functions. But if we just simply want it to happen, it will not happen. If we work for four days and nobody comes to give us anything, our traditional stories tell us, then we have not worked well. We are sweet, just like the doughnut we have been given. That means we have sweet, healthy, and strong thoughts. Our thoughts are in balance.

After some moments of silence while Mama José Gabriel finished his doughnut, he continued:

We do a lot of energetic work. Here it is us who does this work, but who does this work where you live? You must begin by ceasing to criticize others or thinking badly and instead accept what comes. It is important not to be afraid when someone comes to kill you, that would be completely unnecessary. When the moment of death comes, then that's the way it is and you can't change it, and that's exactly why you should not be afraid.

Here in the Sierra, there are many conflicts between the government and the guerrillas. We simply stay in our houses and keep working in our fields and do not harm anyone; we just continue to do what we normally do. Then when the army comes, they greet us and move on. When the guerrillas come, they greet us and they too move on. If someone comes to take profit, then let them come. We do have a chair or a wooden stick here, they can take it if they want. We will not resist them, for this is how we remain on the path of Sezhankua. The Kankuamo wanted to fight for the guerrillas and many of them were killed; we Kogi didn't even think of conflict, and that's why we live in peace. People came with their guns but they had no interest in us, but they were interested in the Kankuamo, even though they are also natives.

We do not criticize. Everyone was given exactly what they need. Some have lost access to it, but that is not a reason to criticize them, because they do have the knowledge, it is just hidden.

The Kogi do not change their behavior in any way in the face of an apparent danger or threat but continue to keep doing what they normally do. They do their tasks and go to their fields or to their gardens. It is not a matter of denying or ignoring a situation but of meeting an event in a neutral and clear manner. They explain that by doing so, they do not become a target for violence.

I was already familiar with this idea from Eastern philosophy, yet nowhere had I ever seen an Indigenous people living and dealing with adversity in such a radical way as the Kogi do in the Sierra. To better understand, it is important to clarify the historical context.

HISTORY OF THE REGION

Between the years 1958 and 2012, internal conflict within Colombia claimed over two hundred thousand victims and led to almost 5.7 million displaced people. For a long time there have been various stakeholders and violent perpetrators: several guerrilla groups, the paramilitary, many drug cartels and mafia organizations, as well as the Colombian government, all of whom are not exactly known for their careful approach or gentle methods. The Kankuamo actually experienced significantly more fatalities than the Kogi, of which almost none have died. This is a remarkable achievement in such a violent country.

The principle of acceptance, as the Kogi practice it, is the foundation of how they deal with all the challenges brought to them by the Younger Brothers and Sisters. Toward the end of the last century, the Kogi experienced a period of cultural disillusion that lasted for several decades. They suffered from the influences of colonization, land theft, and mental health decay through the assimilation attempts of Christian missionaries. Another factor was the attempt of the government to integrate the Kogi into Colombian society by building and funding schools for the children. All this affected their way of life and resulted in a weakening of their traditions, which sometimes even manifested as illness and food shortages.

The Mamos' response to these threats was to return to, and strengthen, their original culture, including that of the principle of acceptance. Even though they have not succumbed to the Colombian government's attempts at controlling them, they have nevertheless accepted the presence of some schools and have used them to their own advantage: now only the Kogi children who are born with the task of being in contact with the Younger Brothers and Sisters are sent to the Colombian schools.

It is absolutely essential that all of their achievements are based on upholding their unique way of life. The restrengthening of their traditional culture prevents the Kogi, almost completely, from mixing with or being conditioned by influences of the modern world. Their thoughts and unique view of the world is their strength.

When we take care of the world, the world will take care of us. The Kogi are convinced of that. Good and balanced thoughts are thoughts that connect people to the world. An essential part of this is acceptance. The Kogi approach life as something that is always *for* them and not *against* them. Therefore, even adverse situations are seen as a gift of life and accepted. This approach in life allows them to be open to all possibilities and strengthens their ability to act well, especially in unfavorable circumstances.

CHAPTER EIGHT

How Will We Make Sure That Everyone Is Well?

Arhuaco Mama Wintukua Kunchanawingumu shares:

To be healthy means to have peace and quiet, to be joyous, to be well in our thoughts, in our body, and in harmony with the land that is our territory. Life has a natural law. We are not alive to live by the way we want; we are alive to fulfill the task that has been given to us as a community and to each and every one of us. This is life.

Health means to live well. This is the best medicine for your body. On the path through life it is not about walking just for the sake of walking. It is important to know where we are going. We natives know where we are going and why we are doing this. Even when we know where we're going, we must join together and look back to reflect to see whether we're going in the right direction.

In contrast, the Younger Brother never seems to pause. He never looks back, he is just looking ahead, regardless of what happens to him. Even if it is not the right way he will complete it, and so the way can lead to his own destruction. To pause in order to correct and adjust and to reexplore the direction is very important. So let's start to consider how our ancestors thought, that's why we get in touch with those thoughts.

However, it is not enough to just examine or order our thoughts and ask for the right way to go; we must also follow our thoughts. Also, the thoughts reveal various tasks and actions that we will put into practice. They determine how you live, how you act, how healthy you are, how your thoughts and your life are, and the way you relate to the world.

Humans have experimented with various forms of social coexistence since the beginning of time. Cultural traditions, organized religion, and political ideologies have shaped various types of social structures, hierarchies, and forms of governance. Even in today's organizational challenges or the whole movement of New Work, these experiments continue.

However, beyond the many surface differences, there are two basic and fundamental forms of social coexistence: community and collective, and what distinguishes the two from each other is the role that the individual plays.

Communities commonly comprise authentic, independent individuals whose fundamental differences are seen as a foundation and a strength that uphold the common vision. There is an emphasis on supporting each other, both consciously and unconsciously, in developing uniqueness and personality, that each individual within the community encourages. Embedded within the community is a deep trust and acknowledgment of the positive synergic effects that naturally arise from this approach, that, in some ways, resembles a balanced ecosystem or a healthy family. The community allows the individual to grow and flourish naturally, and from this the community itself then grows and prospers within greater society.

Collectives, on the other hand, are more often based on an ideology that must be adhered to, that tends to disperse the uniqueness of an individual and attempts to assimilate them into the collective whole. This in turn creates a disconnect from oneself and aligns the individual with deriving their identity from belonging to a group and not from their own personal independence. There are countless examples of this throughout history, such as totalitarian states, sects, and political ideologies. However, there are also far less obvious examples today of unquestioned compliancy within groups of society, such as many of our modern subcultures, within fashion trends, momentary social media hypes, and varying peer groups. In collectives, which often falsely purport to be based on individualism, the human need for belonging is ultimately greater than the will to strengthen one's own uniqueness.

From this viewpoint, the Kogi are a community, even though the uniformity of their white clothing may suggest otherwise. The level of autonomy and independence that each Kogi family receives within their culture is far higher than within ours. Each family feeds and clothes themselves, as well as building their own house, without any external dependency or intervention from the group as a whole. Their culture is, within itself, self-sustaining and based on self-reliance and does not necessitate any central governance, beyond the advisory capacity of the Mamos and Sakas. There is no tax, no fines, and within the Kogi territory virtually no crime.

In the community of the Kogi everyone is a participant and there are no uninvolved observers. On the other hand, in collective hierarchies, leaders are elected and assume responsibility for the collective, while the people who are subordinate to them merely have to follow instructions. Having structure and coordination within a community is sometimes also desirable and has its benefits; however, as is true in the case of the Kogi, it is first imperative that each individual is emotionally and mentally balanced so the community as a whole may be balanced. Of course, the Kogi don't just live based on individual will or preference. They spend a lot of time synchronizing everyone's thoughts and perspective in the *nuhué*, the ceremonial house.

There are also the local leaders and traditional authorities in charge of the different villages, and people must ask their permission if they want to do anything out of the ordinary there. There is also a representative who works on external matters, with the outside world, primarily in relation to the Colombian government—he is called the *cabildo gobernador*—and in other traditions may be seen as a chief. All of this is relatively new to the Kogi and has formed a type of hierarchy within representation. However, the basic cultural setup of valuing irreplaceable individuality as the basis of a strong community ensures their survival until today.

The Kogi live in a thriving community that has existed for well over four thousand years—comparable examples of which, in our current world state, are few and far between, and ever-increasingly diminishing. So just how have they maintained such a vibrant, durable, and capable community for so long, despite facing the same fundamental questions and challenges of coexistence as the rest of humanity?

In spite of some large-scale attempts to eradicate their traditions and peoples, the Kogi have survived for millennia, and especially when combined with their resolute practice of nonviolence and focus on emotional balance, it is extraordinary that what remains in strength is the exact thing that has allowed them to achieve such a supreme achievement—their culture.

WHY ARGUE?

The Kogi believe that, for example, arguing is completely absurd as each of us, and everything that exists, resides within the same internal space. So if you reject or contradict some-one's thoughts or feelings, and that of say, a plant or a bee, you are essentially dismissing and undermining the whole rather than the individual. The resulting impact is far-reaching.

There are many instances in human history where whole societies and cultures have completely disappeared within a relatively short period of time. The question remains, just why were these societies and cultures unable to recognize early signs of their own decline?

The answers are complex. It is well assumed that focus was on the short term, so decisions and actions that were made worked for a while, but in the long term proved fatal. In society today it is easy to find many similar examples of cultures becoming persecuted and fragmented, on a clear path to destruction, if we are willing to take a deeper look.

The Kogi do not discard the value and impact of an individual on society and the whole planet; put more simply, for them there is no separation whatsoever between the two.

Mama Pedro Juan Noevita explains:

If we support each other every day, we become stronger. Also, in the firmament too, the stars are in communities. That's why we see many stars up there and not only the moon. They've gathered and they live together. We align our communal life with the rhythms of the universe to order and organize our community.

We work together from January to May or June. Then between June and December, it is *Nabusizha,* and communal work shifts to individual work. It is the time to put seeds in the earth, receive the *poporro*, to get married, to make *zabiji*. This period, when we return to our individual work, is called Nabusizha. We observe and analyze everything deeply. Half our time we dedicate to community work, and the other half to individual work.

Some people do not want to do the work they have to do; that's why bad thoughts, bad ideas, and bad intentions arise.

Nowadays, the thinking has changed, which is because we have failed to organize ourselves and our thoughts collectively. Some tasks have to be fulfilled communally, while others are done individually, but this does not mean that we will follow a different path. We arrange new types of meetings in order to reach a common understanding, to unite and to progress as a community.

The meetings and traditional rituals of the Kogi are all organized in accordance with key dates of the annual cycles of nature in order to maintain a basic community structure that incorporates not only people, but also plants, animals, and even the wider cosmos. This is an essential part of the Kogi's fundamental cultural practices, as for them all life follows a rhythm.

Every fortnight each person goes to clarify and cleanse their thoughts with the Mamos and Sakas. The Kogi view wisdom as the preeminent measure of value, one that truly designates status and determines the authority in decision-making. Genuine wisdom can only be found through experience and integrity

and enables holistic, widespread solutions to complex challenges; genuine wisdom is an external representation of inner
order.

WHAT IS A WRONG?

In our world it is a criminal offense to flout the law, but in the
world of the Kogi it is not the person who causes an offense
who is considered guilty, but the community, the family, or
even the Mamos, and they instead seek to find the reasons
why that person has been unsuccessful in respecting the order.

There is no victim or perpetrator, there are simply only
those who are involved in the event. Responsibility often falls
on those who inadequately taught the supposed "offender"
correctly, and who consequently failed in ensuring long-term
balance within the community and in the natural order of
things.[1] The act, or offense, is seen as a symptom of the neglect and lack of guidance the person was given. There can
also be far more complex situations that may involve larger
groups of people or several generations, yet the goal is always
the same: it is not about finding justice in the sense of right
and wrong, perpetrator and victim, punishment and compensation, the goal is to restore the order and the natural way of
things, which at the same time acts as prevention for any
future problematic situations.

The Kogi acknowledge both social and individual offenses.
Offenses can be general, such as of a sexual or religious nature,
and can also be connected to the belongings or property of

others. Any kind of physical destruction is an offense for the Kogi, as well as the failure to adhere to the family structure. Another offense is the betrayal of sharing secrets by passing on knowledge to other peoples who have lost their own original wisdom. (Even for us, they are only passing on a small portion of their wisdom.) For the Kogi, the thought of the crime and the crime itself are of equal measure; therefore, the very thought of stealing yucca is the same as the act performed.

Confessions of offenses are held in public, and all participants comment on the incident during the assembly. As a consequence, those involved might have to learn specific cultural legends and stories that show parallels to their behavior and that potentially describe a resolution of the incidents and thus restore balance. The Kogi do not see this as a punishment, and neither would we.

For us, a sense of justice, if violence or coercion has been committed, is attached to whether the suspected perpetrator has committed an act that is deemed legal or illegal. In our society, civil servants elected by the people are allowed to collect taxes or lock people up in prison, but if individual nonelected people do the same, it is robbery or deprivation of freedom. Our idea of punishment was brought to the Sierra Nevada by the Catholic Church and its attempt to Christianize the four peoples of the Sierra. This attempt failed, with the exception of the Kankuamo.

Kogi Arregocés Conchacala, who worked with the government, shares:

The name for justice is *sezhagawi*, and it describes the order of the people and the relationship between the society and the territory. Performing and achieving justice is called *kwalama*. The traditional authorities of every community hold this responsibility. Approximately in around the year 1890, the Younger Brothers began to enforce their government here. The administrative inspectors sent by the Younger Brothers determined the punishments, the sanctions, the laws, the concept of exchanging earth for money, censorship, the prison system, and the church for our communities: one had to get baptized, and also married in a certain way, there were rules on how we must live together, and many natives obeyed. So our *kwalama* [food security] has been weakened. They wanted us to obey their artificial laws. This is how disorder began to spread. The *esuamas* [places of power] became weaker, and we lost the possibility of maintaining our culture the way we used to, and the resistance against our culture grew. Our authorities then removed with great determination the church that had established itself in San Antonio. And today we are talking about the *esuamas* again.

We face several difficulties that the government schools of the last 120 years have brought to us. To resolve this, we need patience. There were even places where the Younger Brothers demanded fines to be paid and imposed various penalties. The problems of artificial justice remained here in the form of negative thoughts that the Younger Brother brought to us. It was a deprivation. That is why some today even think that the way to solve a problem is to punish a woman outside of the nuhué. This was the way the Younger Brothers' inspectors approached the problems: they would beat the women there. Our big problem is that many children saw this and grew up with it. Today, when we say that we will visit different sacred places to order our thoughts and actions, it is not easy for some natives to participate, because they are now used to punishments. You can see from this example how bad thoughts take root.

THE KOGI SAY THAT THE BASIS FOR A MAJORITY OF SOCIAL problems is people becoming dependent on other people in an unbalanced way. To me, this makes sense, because dependency instantly creates a hierarchy: the one with and the one without. To become dependent on another person gives someone else control over your needs and the choices you make. This can more often than not lead to resentment, frustration, and low self-esteem.

The way the Kogi live simply avoids such a situation, and in many ways the Kogi community resembles nature: each person is capable of surviving on their own and yet is wholly dependent on the ecosystem of which they are a part.

The Indigenous of the Sierra Nevada have on the one hand much more autonomy than we do, while on the other, and perhaps even because of this, have a more functioning community.

However, leaving the community to follow one's own "freedom" is not a possibility. Unlike with many other Indigenous peoples, very few Kogi have left their communities to live in the city. Leaving one's community is seen as abandoning one's cosmic purpose, which, according to the Kogi, is split between one's own individual purpose and the community purpose.

Being a Kogi is not an identity of choice, such as being a yogi (in the modern Western sense), but rather being a representation of a specific cosmic blueprint that is deeply interwoven into the community, and the land of origin. The question arises, what does that means for modern societies, such as those in Europe and North America, with centuries of migration, with some societies being as much as 99 percent derived from settlers, and how does it shape the way these societies live and perceive the world?

The Kogi can't, and won't, answer that; it is a journey of understanding we must find out for ourselves.

EXPRESSING ONESELF TO THE COMMUNITY

A key component that has afforded the Kogi the cultural resilience that they have is their approach to dealing with their thoughts and emotions, which are seen as barely indistinguishable, and which they neither suppress nor ignore. If emotions are suppressed, they can internally manifest and cause destruction inwardly through disease and a lack of self-connection, and if they are uncontrollably expressed, they can hurt those around us and distance us from society. Even though we are aware of this, in our culture, we are more often than not compelled to control and suppress certain emotions, at least in public, but often also within our family. Outbursts are reprimanded, sanctioned, and in the worst cases, the person at fault is sometimes punished with exclusion from the community.

The Kogi approach this topic in a very different way from us and are fully aware of the severity of neglecting it. Negative emotions are seen as an intrinsic and inseparable component of life; therefore, the Kogi do not distinguish between positive and negative emotions. The pivotal practice they use in dealing with emotions and thoughts is called *confesar*, a word based on the Spanish vocabulary of the missionaries, literally translated it means "to confess."

To be clear, this has absolutely nothing to do with the confessional practice seen in Catholicism. In the Kogi language it is called *alúnayiwási*, with the most appropriate English translation being "to express something," and it means to carefully express one's thoughts and emotions in the circle of the community, and to the Mamos and Sakas. They will express everything, both positive and negative, that they have thought, felt, said, or experienced to themselves or to others. The community creates a space, which is extremely effective for personal and communal confession; there is no judgment, no exclusion, and no blaming or shaming. The confession is an essential, regularly conducted ritual in their culture and is a natural part of life for all Kogi above a certain age. One of its key focus points is, as the Kogi say, to fully inhabit one's own home, which means to be fully balanced in one's own physical and subtle body, as well as in the community.

Ade Wiwa Mama Ramon Gil Barros explains:

We already organize ourselves from the moment of conception, at birth, or when we receive the *poporro* and during marriage. We order our vision, our hearing, our words, our heart, and the masculine and feminine principles. We do *ruamashka*,[2] that is, the ordering of *ruama*. If someone just focuses on their internal order, they don't grow, and if they just focus on the external order, then also they do not grow. In order to grow one must be aligned both internally and externally. *Ruama* is within and is guiding. It is the essence, the clarity, the strength, the knowledge, and the wisdom. When *ruama* leaves, the body is not of any use anymore.

When we reside in one place yet our thoughts are in another, even if we are remembering or dreaming and not present in the moment, then what we hold within, our *ruama*, will leave us. If we are always living externally, and sometimes spend years out there, then we lose ourselves. In this way snakes or other creatures can enter our house, vermin are born, and everything comes to an end. However, if we inhabit our home and maintain it well, we will stay alive and will not easily deteriorate. When we practice *ruamashka*, the ordering of our self, then it is guiding and leading us. We practice *ruamashka* so that *ruama* does not escape from our nuhué, so that it does not leave us, our body.

OUR BODIES REVEAL US

The inner order always has an outer expression. The Kogi say that by looking at our physical bodies we can see how we think internally. Intrinsically, the focal point for bringing the human body into balance is in the center, where the Japanese locate the Hara. The modern ideal of masculine beauty, of broad muscular shoulders, reveals a shift of that center to the upper part of the body. For the Kogi, this is a sign that we have put greater importance on our own vanity than what is intrinsic and natural in life, and we are no longer in balance.

Mama Ramon Gil Barros continues:

Our personal ordering begins before birth and lasts until death. It is most intense in the four most important stages of life: the birth, the onset of puberty, and in adult life, in marriage and death. In these four moments we order our behavior in material life, in our body, in our thoughts, in our desires, and in the memories of our life. The personal fulfillment of the communal and territorial order begins with the ceremony of birth: we call it *gonatushi*. For nine days and nine nights, the Mamos present the child to the elements of nature that they will encounter in their life: the water, the fire, the food, the animals, the people, and the trees. The mother and father also have to inwardly cleanse and balance everything, that is why it takes nine nights and nine days. From the moment of conception onward we work with the couple through *alúnayiwási*, in everything they think, do, and dream.

The confession takes place either on a regular basis or whenever it is needed. Every big change, major decision, or new activity is reviewed with a Mamo or Saka. Expectant parents go for regular consultations. For all other Kogi individual confessions occur before every trip, harvest, construction of a new house, or important agricultural activity. The person will approach a Mamo or a Saka with the request to confess and then in the ensuing conversation will be guided and asked questions about all areas of their life. The awareness that is always present is that everything that has happened in Aluna, that is, in thought, has, with the exact same effect, happened in the physical world. The Mamo or Saka will ask:

> What did you think when you heard that your wife
> was pregnant?
> Did you remember the trees and did you treat them
> with respect?
> Did you talk badly about your neighbor?

Beside individual confessions, men and women hold separate debates in larger groups, where each person voices all the positive and negative things they have either said or thought about other people or in the activities they have done. Everyone in the group remains silent and listens to the words without judgment. Afterward the Mamos or Sakas give their advice, often in the form of parables from the myths of the Kogi. Oftentimes the thoughts that are expressed are collected in small cotton threads and are then taken to an appropriate sacred place.

Confessions are a way of preventing illness, social conflicts, marital disputes, crimes, and social decay. The possibility of expressing what is going on internally for a person without being judged can be of huge comfort and can prevent conflict before it can arise.

Nevertheless, the Kogi do differentiate between positive and negative thoughts but without shaming the individual. If, for example, someone thinks of stealing something from someone, this is seen as a negative thought, and the thought is then returned to the fathers and mothers of theft. The one who had the thought is responsible for allowing the thought to happen but bears no guilt for thinking it.

It is important to understand that the returning of thoughts is not an isolated, purely mental act, but is also an act of energetically relinquishing the thoughts. This distinction is essential, as merely acting with the mind to let go of something only satisfies one part and in essence suppresses the energetic components of a person. Every thought has a corresponding place to be taken, as does every energetic charge.

Mama Bernardo Simungama Mamatacan speaks:

When we confess, the Zhatukwa determines what we will talk about. Sometimes we only talk about the negative because we think and do a lot of negative things. The woman absorbs everything, everything positive and everything negative. That is why we regularly cleanse our thoughts. We also talk about the positive, because once the house is cleaned, we start to decorate it to make it beautiful. If a house is very ugly, you will not feel comfortable there, even if it is clean. Therefore, it is important to express the positive thoughts and to decorate the house nicely. A new life is always a new nuhué, a new ceremonial house. When we, as parents, cleanse ourselves mentally, we support our child. We put these thoughts in cotton, as we ourselves also dress in cotton.

Thoughts are often in competition with other thoughts. There are always positive and negative thoughts on every topic. The Mamos also know exactly which of our ancestors at the origin of the world thought which thoughts on which topic first. They know the stories. For example, Nuanase was also a Mamo, but a negative one. He turned against the community and did not work well for life. He even turned against all of humanity. The Mamos know where these thoughts come from.

When we make a confession, the Mamos take the thoughts that are not our own and give them back as nourishment to those from whom they came. Because these thoughts belong to them and not to us. But first we have to accept them and become the thoughts; in this way we become freed from them. These thoughts are released and returned. That is the work that the Mamos do. They call on the creator of the thought and give back what belongs to them. Sometimes we also call the originators of the positive thoughts, but not always.

The Mamos do not always bring the thoughts back immediately. Sometimes they first gather the thoughts in one place and announce them in the spirit realm to the owner of the thought to let them know that they will come to return the thoughts soon. Once they have gathered quite a lot of thoughts, they then visit the corresponding place and give back all those thoughts to the ones to whom they belong. Sometimes the Zhatukwa tells us that we can just convey the thoughts in the spiritual realm and that we do not need to physically visit the place. Often, however, we go there. There are also times when the person who had the thoughts has to visit the place himself from where the thoughts have come, and return them with the help of a Mamo. If one neglects to do this, the thoughts might not be accepted and taken back as nourishment.

Life is all about relationships, and they determine our successes and failures. In the nuhué, the ceremonial house, children of a certain age explore the different energies and emotions within themselves and experience the balancing quality of the nuhué. It is the place where many confessions happen. This social practice instills such a profound trust and sense of belonging that the Kogi are able to live without social sanctions and hierarchies of power. If the relationship between people is not given care and attention, then leadership is required to prevent injustice and murder, but when there is agreement on a social and emotional level, people can live in real relationships with each other and with all of life.

The Kogi are living proof that there is no innate need for violence among human beings, and that it is far from being part of so-called human nature, as is often claimed in Western writing. They are living embodiments of this truth and fact. They have lived for centuries in peace with their neighboring peoples and wider society, despite the countless attempts to destabilize and constrain them. From children to adults, all Kogi explore the emotions they hold within themselves and follow the social practice of cleansing those emotions in both a mental and energetic way. This instills a profound trust and sense of belonging among the community and ensures they are able to live without hierarchies or social sanctions. Each person is responsible for their own emotions. This builds a foundational structure that everyone can rely on, and the relationships within that community, or society, can flourish.

Can we look at our own lives and see where we can open dialogues about our emotions, both good and bad, and how we can best accept them to restore internal balance?

Can we attempt some form of this in our families, communities, or businesses?

CHAPTER NINE

How Will We Live Another Eighty Thousand Years?

I was staring at the ice-cold crystal clear water of the river, watching how the rushing water formed pathways downstream, veins of liquid life-giving nourishment to each place it touched. I wondered how long it would be before, if ever, I would stand here again. It was the day of our departure. I soaked in my surroundings; every inch of it had become so familiar, it had been my entire world for months now. This place had grown so dear to me, I felt at home on the land, and among the people. I had found a place so full of peace, yet in the same breath, so vibrantly alive. In the pit of my stomach a churning thought was slowly rising—what would it be like to return home, to so-called civilization?

I felt changed on all levels, living with the Kogi had altered me and expanded so many of my perspectives. I had lived a pristine way of life in the Sierra; no exhaust fumes, no artificial noises, no hustle and bustle, all had been replaced with fresh water, fresh air, and the purest of food. While my stay was beyond gratifying, I also knew that I clearly did not belong in the Sierra, it was not my place. I am *de allá*, from the other side of the Atlantic, and not *de aquí*, from the Sierra, as Mama José Gabriel would say. The strongest indicator being the sandfly bites that I simply could not get used to.

We quietly strolled one last time with Mama José Gabriel through his garden, and he began to speak:

People will say that they like the book and that they will respect the sacred places. We have talked a lot. We are writing this book and it will change the thoughts. That's how we will reconnect with our essence and to what has been left to us, also to *zhigoneshi* and to the One Thought. In the One Thought we live well, we order well and organize ourselves well. We do this and in a few years we will see that there will be better water again, that people treat each other better and that the food is better. We will even live happier. We will take care of these thoughts. How are we going to do that?

After writing the book, you will start an academy for these things. When the Elder and the Younger Brothers talk to each other, then something will change. We have to think together! It is not enough for human beings to examine the thoughts and simply say, "Ah, that is beautiful thinking." No, they must live the thoughts and become the thoughts themselves. That is what you will teach in the academy. When the Younger Brothers and the Elder Brothers accomplish their work in the One Thought, then we will live for another eighty thousand years.

We walked on in silence. Eighty thousand years is a long time, I kept thinking, well over 2,600 generations! Was it naive or impressive for Mama José Gabriel to even contemplate humanity lasting another eighty thousand years in the face of our various pressing global challenges that are fed to us on a constant basis through media and technology? Yet, as he spoke I did not sense hope, I sensed a complete focus and assuredness. It did not simply feel as if he was giving me his opinion, it felt as if he was speaking the words of all the Mamos, of all their ancestors and the thousands of Zhatukwa consultations he had been a part of.

So, just what is required for humanity to flourish on this planet for another eighty thousand years? Throughout my stay, the Kogi kept repeating the importance of us reinstating life force back into the center of our thoughts and actions. This sounds almost achievable within the world of the Kogi, but how do I implement it once I'm back home, in the modern, chaotic world? Even after experiencing how they embrace and maintain life force in every aspect of their lives, and have done so for over four thousand years, it's difficult to know where to begin. How do we integrate such a deep and vast respect for life, and the world of spirit, when modern society is dominated by technology and false progress based on artificial solutions? What does a society based on the innate principles of life even look like? What does an economy operating according to these principles look like? What do we need to realize and shift to become the humans who are capable of achieving this?

Mama Bernardo Mascote Zarabata shares:

We will all find our origins again. Progress does not mean to restore the beginning, but to preserve the origin and keep it alive. When we pass the knowledge of our wise ancestors to our children, we live well. The most important thing is to respect nature and to stop building large industrial projects that harm nature. The Mother does not want to end our human existence, but if we do not follow her way again, then she has no other option. The artificial laws and norms that we are inventing today harm the Mother. Things are not right simply because they have been made legal by a human-made law. What makes us think that we can issue laws ourselves? Why is it that we believe that we can invent a law? These laws are used to legalize mines but that doesn't make them right. It is still very harmful to build mines, even if they have a license and are approved by governments.

We follow the original principles. They show us how we can

guard all things and take care of everything. The water, the trees, the stones speak to us about the laws of the Mother every day. We just listen to them. These laws have never been changed; Jaba Sé created them in the beginning and they have existed ever since. The politicians invent laws in order to be reelected. It is only with invented laws that it is possible to destroy nature. If we regain our original knowledge, then the Mother will listen to us again and help us again. If not, then we continue to follow the path of self-destruction.

The most important thing for a human being is to know the principles of the origin and the principles of life, this is how we create a very clear and bright future. If we do something or learn something that is for the good of the Mother, then she will strongly support it. If you want to regain your knowledge and you are not sure if you are doing it well, just pay attention to whether the Mother is helping you or not. It is essential that you reclaim your traditional medicine, the spiritual medicine. Chemical medicine is harmful to the body even if you won't see the effects until much later. Natural medicine comes from nature and it gives us everything we need. Nature also has a mother, her name is Jaba Sé. She also supports us.

We Kogi are also here to support you to find and restore your roots again, but it is up to you to bring back your own knowledge by constantly having your thoughts focused on it and being receptive to it. This is very important. Then the Mother will support you to go where you are supposed to go.

Where will you find the knowledge again? It lies hidden in

your earth. Your ancestors lived there and the knowledge is connected to these places. That is where you will find it. It resides in the trees, in the stones, in the water. We still know these things, we still live according to them. We humans think that we support each other by helping ourselves to remember our knowledge, but actually we support the earth herself, that is why she will support us.

The Kogi do not have ready-made solutions for us. What they have is their own culture, their own specific contexts, their own ecosystem, their own language, and their own understanding of the world. It would be impossible for us to become a Kogi, and we should not want to either. Nor should we all travel to the Sierra to learn from them.

But what the Kogi can offer us in vast quantities is inspiration for how to reconnect to our innate memory and cultivate a lasting, meaningful, and reciprocal relationship with ourselves, our communities, and nature. The Kogi can act as guides, giving us directions and indicators of where to look, but they cannot tell us what to see when we embark on the journey. That we must learn for ourselves.

We could take the perspective that the Kogi, and many other Indigenous groups, embody—a "first way." They have all largely—with their own specific approaches—lived for thousands of years in deep relation and communion with nature, with her laws and her principles. Sometimes because they chose to, and sometimes because they simply had to. For us, it would be absurd to attempt to readapt to a traditional Indig-enous lifestyle, least of all considering the strenuous physical challenges and often harsh way of life that many of us would not want to swap our relative comfort for. The "second way" is our modern world. In many ways it is a derivative of the first way, being filled with "second thoughts and ideas" that are mostly detached from nature and disconnected from the re-newal of life, yet they remain successful in producing techno-logical and reductionist material advancements.

It is clear, without a doubt, that we cannot continue in the way we are going—our current global model and system is destroying the planet on an unthinkable scale, creating huge devastation and social misalignment. As it seems impossible for us to return to the first way, and the second way is cur-rently destroying humanity, it feels the only open way is to explore and create a "third way."

The third way consists of taking the most potent and ap-plicable aspects from the previous ways and creating a living future that is self-reliant and self-renewing. These aspects are not just thoughts, technologies, or social practices; they are first and foremost perspectives, narratives, relationships, or in short, consciousness. However, the third way doesn't yet exist, and not even the Kogi know for certain what such a world would look like. As Mama Bernardo Mascote Zarabata said, it begins with an honest dialogue and real contact between the Elder and Younger Brothers and Sisters.

If we, the Younger Brothers and Sisters, become dedicated

and willing enough with our intent to listen, to learn outside of our fixed ideas and perspectives, to want to live within a world that is just and prosperous, if we are true to our innate inner nature, and if we begin the process of removing ourselves from the center of everything, then I am certain change is possible, and a fertile ground for the third way to emerge from can be created. Of course, it is imperative that the same level of openness for cooperation and co-creation applies to the Elder Brothers as well.

EXPLORING A REGENERATIVE WORLD

In our current age, the world is governed by just a handful of main factors: technology, business, agriculture and, more recently, changes to the ecological biosphere and climate. This affects us on both an individual and a societal level. What lies behind all of these factors is the question of relationship: How do we design and establish our relationships, and are they nourishing and beneficial to us?

When I asked the Kogi about their view on the priority we place on technology, they emphasized that it is in fact the task of the Younger Brothers and Sisters to work with technology and machines. However, they stressed, this should be a technology that functions in harmony with the earth and its many living principles, and that does not work against them, as is often the case today. I began to seriously consider what a technology that works in harmony with nature might look like. The Mamos reassured me, and explained, that technologies based on the principles of natural order and life force are absolutely possible and that first attempts have already been made; now it is mainly a question of supporting these attempts. It is worth noting that Mama Bernardo, who explained this to me, had never left the Sierra Nevada, yet he explained that he had seen all of this in Aluna.

After I had arrived back home I began to investigate what the Mamo might have meant by the term *living technology*. He had indicated that in some way it already exists. Many of the things that first came to mind, such as electric cars, solar energy, soy-based meat substitutes, and so on, proved at a closer look to be less sustainable than I first assumed, especially in regard to their production and implementation. Technological development and economic growth are generally focused on the aim of increasing convenience in our lives. Whether it is finding ways of transportation that allow us to get from A to B quicker, in a more cost-effective way and without being exposed to climatic interference, or washing machines that relieve us of the tedious and strenuous task of hand washing, or smartphones that give us ubiquitous access to enormous amounts of information via the internet. We perceive these things as enriching because they relieve us of hard work and ideally give us more freedom and possibilities. Yet, what do we lose in the process? In a search for greater ease and comfort, what knowledge is being lost, and do we at any point take life, and nonhuman living beings, into consideration while creating new systems?

Biomimicry is an approach that could potentially be a first step toward a living technology. Biomimicry is a science that is oriented toward nature and attempts to imitate it to solve problems in our modern world.[1] The fundamental question for every arising problem is, "What would nature do in this situation?" Biomimicry is divided into three aspects: first is the emulation of nature in the functionality of a product or design; second is the method of production and its processes; and third, the way something interacts within its ecosystem.

What is the source of all great inventions? Leonardo da Vinci, one of the greatest geniuses in human history, said that all his numerous inventions came from nature. He spent hours each day observing and sketching the anatomy and flight patterns of birds, the growth habits of plants, and the behavior of water. Based on these observations he compiled a wealth of design sketches that were imitations of nature. Why did he do this? Quite simply, because nature is far more ingenious than man, and has already solved an overabundance of technical problems in the most optimal way: binding things together, making them waterproof yet breathable, filtering salt from seawater, converting carbon dioxide into oxygen, producing extremely strong yet flexible materials, and much, much more. The examples are countless. The solutions that nature provides are almost incomprehensible, not only in terms of the materials it produces but also in terms of its energy efficiency and sustainability. It does all this not as a by-product, but because it is natural for nature to act this way.

A natural solution, by definition, is something that merges perfectly into the cycles and rhythms of life because it is formed of nature and the universal guiding principles. Therefore, it stands to reason that nature provides solutions for our problems, and that it must be observed in its own habitat as closely as possible, without the interference or manipulation of modern humans, so we may have the opportunity to witness and apply its principles to our own way of living.

For this to happen, two things are vital: We must first understand that we cannot truly observe nature, as that suggests there are two parties, the one that observes and the one that is being observed. In reality, there is no separation or division between all living beings and nature—so it is imperative that we step into nature with a desire to absorb what exists, as well as trying to analyze or quantify, but with total appreciation and awareness.

Second, it is vital that we do not simply read and learn about nature from books or other people, but that we each experience nature in a direct way, that we are touched by the force and mystery of nature, that we can feel her with our hands, that we can hear her with our own ears, in whatever capacity is available to us, regardless of where we live.

We are all a part of nature—this fact can be argued less and less—so this is what makes it possible for us to feel nature, because we are it, so we can feel it within our own body, and it could be said that this is where truth, or our innate "knowing," resides, in a connection to one's own body and connecting to the feelings our body produces. So when we surround ourselves with the natural living world and feel what our body is producing, in sensation, beyond our analytical thoughts, we are able to connect to the innate truth within our essence.

LET'S LOOK AT ONE EXAMPLE OF HOW WE CAN UNDERSTAND A Kogi perspective on an issue key to our survival. The importance, and the vital role, of soil is commonly overlooked when we discuss the well-being of the planet and the survival of humanity. More and more studies indicate that there is global soil erosion occurring with alarming acceleration. Some even claim that, on average, we only have about sixty harvests left before our soil is totally depleted. In most climate zones, that is just sixty years.[2] Erosion develops when soil essentially turns into dust and the earth is left barren, without nutrients or microorganisms. So how, and why, does this occur?

One of the main reasons is our current system of, and the processes within, large-scale industrial agriculture. The combination of pesticides, fertilizers, and tilling may boost efficiency in the short term, but the long-term effects are wide-reaching and devastating. Fundamentally, the soil becomes heavily degraded and is stripped of any, and all, "livingness." Large-scale tilling, for example, stresses and mixes the earth in such a way that the subtle balance of microorganisms, including bacteria and mycorrhiza networks of fungi that are indispensable for soil health, are destroyed. Artificial fertilizers cause there to be far fewer nutrients transported to the plants, and the amounts used will constantly need to be increased as the soil continually degrades, causing toxins to run off and pollute the waters. After the harvest, if the soil is left without crop cover or mulch, the wind blows the fragile soil away and turns it into huge dust clouds, and thus the process of desertification begins. This leads to a highly significant decrease in the natural capacity of the soil to retain rainwater, which causes rain to flow across the surface of the terrain into the rivers, taking any existent soil nutrients with it, and causing devastating flooding and severe damage to nature, while also gradually affecting the rise of sea levels.

The consequent effects of soil degradation are vast and affect us all, and it is a topic given far too little attention among mainstream media. It is also yet another example of the effects and ramifications of thoughts that are detached from nature and her principles. Even though our global net wealth in the world is increasing rapidly, in terms of both financial and material assets, the quality of the soil, and thus our most vital and most precious asset for the continuation of life on this planet, is decreasing exponentially.

The severe deterioration of the earth's soil also alludes to another major, yet subtle, point for reflection: How do we expect to live within a balanced and harmonious culture if the very foundation we live on, and rely on for our existence, is dead? What does it say about our inner state of being if we are more consumed by the ease with which we meet artificially set targets than by the health of what sustains us and the Great Mother?

Soil is the perfect metaphor for a healthy culture—and for a balanced and well-maintained civilization—without it, we can only be a dead culture, in every sense of the word.

We can look to restore this fundamental aspect of our world.

NATURE HAS ANSWERS

In our daily actions and thoughts, and attempts to create solutions and innovations, it seems as if we often forget that something other than human beings were here long before us and first created many things we are trying to achieve. We are not the first to process cellulose, to produce paper, to create optimal surfaces, to impregnate material and to heat or cool structures. We are not the first to build houses for our children. Yet it seems that in our modern culture we have forgotten what we used to know: that we live in a universe that is innately alive and full of extraordinary brilliance, that within our planet we are surrounded by genius, and that we therefore also naturally carry that same genius within us.

The importance of this simple concept is enormous. Taking it even a step further, by honoring and revering nature's waters as a living counterpart, one that allows us to live, we create an unintentional consequence of honoring ourselves, and all beings, as we all contain water within us. If the majority of the world lived with this awareness, the impact can scarcely, today, even be imagined, and the unquantifiable potential and capacity for renewal is immeasurable.

We are now, in some small ways, beginning to remember this again, and that is exactly the remembering the Kogi speak of. We are beginning to live the way all other living beings do, and in doing so, we begin again to acknowledge nature as our ally and kin.

CO_2 EMISSIONS ARE ONE OF THE BIGGEST PROBLEMS WE FACE in our modern world, but this is not a problem for all of us: for plants, carbon dioxide is not a strong poison. Instead, plants transform the CO_2 into food, that is, long starch and glucose chains, and thereby release the oxygen we need to breathe. Scientists at Cornell University have changed their perspective in response to this observation and now consider CO_2 as a potential gift. They discovered how to produce polycarbonate, a biodegradable plastic, from carbon dioxide, and so the problem of CO_2 has become part of the solution.[3]

In the world of plants, carbon dioxide is also an essential component of corals. A cement plant in the United States has imitated the ingenuity of the coral reefs, and now its engineers are using carbon dioxide as a fundamental part in the production of cement. Normally, in the production of a ton of cement, one ton of CO_2 is produced and emitted, but they have managed to reverse this equation, and now instead, half a ton of CO_2 is stored in every ton of cement produced, and this is thanks to the coral reefs.

Also when it comes to keeping environments free of bacteria, we are not the first to attempt to do so; nature faces the same challenge. For example, the Galapagos shark has no bacteria on its skin, no growths, no barnacles, nothing. Researchers found that the shark has specific denticles on its skin and applied this discovery to the production of Speedo swimsuits, which have been worn by many Olympic record breakers. These denticles contain a specific pattern that prevents bacteria from attaching to them.

The company Sharklet has adapted this pattern and uses it for surfaces in hospitals. Nowadays, many organisms have become resistant to antibiotics or harsh detergents, and more and more people are continuing to fall ill from multiresistant hospital germs; however, with these types of denticle surfaces the transfer of harmful bacteria is prevented from happening.

Everything that is produced by nature is, by definition, biological and also biodegradable. With this in mind, the materials that exist in nature are fascinating and inspiring. A beetle's exoskeleton is composed of chitin, which is waterproof, strong, breathable, and resilient. Chitin creates color through its structure. While a potato chip bag is made of seven layers of material in order to feature the same properties, the beetle requires just one. Nature harnesses a total of only five polymers to make this possible; humans require around 350 polymers to achieve the same result. One of the most necessary steps to imitate the abilities of living creatures is to use the materials according to their intrinsic design and in the quantities we see in nature.

In this we find a direct link to the message the Kogi shared with us. If we were to stop mining hard rock and instead we extracted metal from wastewater, this would go a long way to supporting the Kogi, and the solution already exists. Within nature microbes extract tiny particles of metal from water, and a company in San Francisco has imitated this process and now incorporates the microbes into filters in order to use flowing wastewater for metal extraction.

In our manufacturing process we use, colloquially speaking, a heat, beat, and chemical treatment process. This is the exact opposite of what happens in nature. In many cases, our production processes result in 95 percent waste and less than 5 percent net product of the raw material. The material is heated, hammered under high pressure, and modified with chemicals, creating an artificial outcome. Nature could never, and would never, adopt a practice that involves such a high level of waste. No wonder that we feel threatened by a scarcity of resources.

In contrast, how does life "create"? Through combining information with matter and thus becoming structured. Without this information, matter would have a completely different function. Self-organization happens through interaction and cooperation, and through this process, material is generated. A mussel shell, for example, is a self-constructing material that is twice as resistant as modern high-tech ceramics. The interesting thing, however, is that mussel shells are created in seawater, whereas ceramics are produced in a kiln. The universal intelligence that creates the process of ordering the growth in and around the body of the shell simply happens naturally, while producing ceramics requires a lot of energy in the form of heat.

Applying this process to our technology would mean producing ceramics at room temperature by dipping them into a liquid and lifting them out again, and then allowing the evaporation to order the molecules in the liquid that then fits them together like a puzzle. This is exactly how crystals are formed. What would our world look like if we produced all our hard-surfaced materials in harmony with the principles of nature? For example, we could simply spray the starting material for a solar cell onto a roof and it would organize itself into a layered structure and begin to capture light. These things would happen naturally because it is in their intrinsic nature to do so and not because we have forced them.

From these insights and observations of processes, green chemistry has emerged. Green chemistry is an emerging sector that replaces, and ideally makes obsolete, industrial chemistry by incorporating nature's handbook. The big difference is that within nature only a subset of the elements from the periodic system are used, while in traditional industrial chemistry, all are used, even the most harmful and toxic. It feels important to explore green chemistry in far more depth and discover the elegance and ingenuity of nature, in order to discern exactly which subsets of the periodic table must be used to imitate the miraculous materials nature is composed of and creates.

"When the forest and the city are functionally indistinguishable, then we know we've reached sustainability."[4] Janine Benyus's quote from her book *Biomimicry* highlights the third and most interesting aspect of biomimicry, the aim of integrating the processes of human life that unfold in harmony with natural principles with the living biosphere of the planet without harming it. This relates to the imitation of ecosystems, biotopes, and larger natural contexts themselves. It is this exact aspect that the concept of the Kogi's maintenance of the territory is based on. This understanding stretches far beyond the mere imitation of a material or a production process and even beyond cybernetics. This aspect of biomimicry describes the relationships and interactions of all living beings within an ecosystem and applies it to humans, both in our internal relationships and in our interactions with all of existence.

We may begin to realize, if we rid ourselves of our generational prejudices and our limiting rational mind, that a forest is not just a recreational area or a collection of resources, but a space of enormous living potential. All of nature's solutions are embedded in a larger context, of which humans are also a part: the earth. As Canadian environmental activist David Suzuki states, "The way we see the world shapes the way we treat it. If a mountain is a deity, not a pile of ore; if a river is one of the veins of the land, not potential irrigation water; if a forest is a sacred grove, not timber; if other species are biological kin, not resources; or if the planet is our mother, not an opportunity—then we will treat each other with greater respect. Thus is the challenge, to look at the world from a different perspective."[5]

If, for example, a factory is to be built in a landscape still full of nature, the question arises: What does the factory have to provide in order to ecologically balance the land that will be sacrificed? The answer must factor in the conversion of carbon dioxide into oxygen, purifying the water and air, and securing the habitats for small creatures and birds.

This is, however, our next and most crucial step, because ultimately it is not just about replacing a man-made system with a natural one, but about the reintegration of the human world and modern technology with the principle of innate life force, of which the Kogi speak often. We cannot learn from nature purely through rational analysis, it can only take us so far. To gain a deeper understanding from nature we must engage with an inward connection that also acknowledges the invisible forces and energies at play, which in turn widens our awareness and comprehension of life force. For us, all of this is truly about having a successful relationship with nature. Fundamentally, creativity is borne from a spark of connection with something and is often a collaborative process, whether with a person, a canvas, a natural object, an idea, or even another living creature. It becomes less about imitating specific natural solutions and more about real innovation that is guided by the principles of life. In the long term, true inspirational creativity is only possible when we align our thinking and actions with nature.

Let us imagine for a moment that we want to artificially orchestrate spring. Imagine the staging, the timing, and the coordination. Naturally, spring unfolds without hierarchical rules or guidelines, without five-point plans, shortcuts, life hacks, or climate change protocols, and it happens every year. It would not be possible for us to foresee, control, and manage everything relevant with a computer, no matter how much processing power it has. It is only possible to accomplish this if there is an original inherent balancing dynamic that is maintaining the natural order. The Kogi would call this the One Thought that permeates all life. We can also use these dynamics to our advantage.

When the Kogi say that the Great Mother created everything perfectly at the origin of time, this does not imply that we should not change anything. Change is a naturally occurring action and part of life; besides death, it is the only certainty we will experience. However, it is crucial that we align ourselves with living change, the change that occurs and is governed by universal intelligence, so our thoughts and principles can become living too.

All organisms in nature contribute in some way to the life of their biotopes—their habitat—through their mere existence, while these biotopes in turn support the descendants of the organisms. Increasing fertility is one way this happens. Organisms increase fertility through their life processes and from this, more opportunities for life. Ultimately, this results in an increase or preservation of life force. This is exactly what healthy ecosystems do naturally, and it is also inherent in the culture of the Kogi.

The Kogi's undeterred understanding of the significance of life force encourages them to give back to the earth what they take from it; they do this in both a material way and by acknowledging what the earth has given them, and they do this as naturally as we drink water. Janine Benyus concludes one of her lectures with the words: "Life creates conditions conducive to life. It builds soils, it purifies air, it purifies water, it mixes a cocktail of gases that we all need to live. And it does all this while fulfilling our needs. Nature does not exclude anyone or anything."[6] When the Kogi speak of the One Thought in this context, it is about humans having to integrate themselves back into the natural cycles of life. This is especially important in regard to our modern technology.

Given the dire consequences that are a result of our modern mindset, it is clear to see that we do not know better than the Indigenous peoples whose cultures, like the Kogi, are rooted in a consideration of, and a connection with, the inner principles of nature as a whole. With this in mind, no matter how many machines we create, how many solutions we find, how many concepts we put into action, the knowledge of Indigenous people is always at an advantage, as it is not possible for them to exist without having recognized a deeper connection to their own ecosystem, which provides the foundation for them to exist.

With every species that becomes extinct due to artificial influence, an array of solutions are lost, and with every Indigenous people that disappears, we lose the wisdom of a group of people who have experienced, understood, and supported so many vital and naturally occurring solutions for thousands of years.

It could be argued that throughout history many different cultures and peoples have appeared and disappeared having seemingly little to no effect on humanity as a whole. So why should we care about listening to one more culture, even if our disinterest will almost certainly lead to their demise? Perhaps simply allowing an ancient culture to slowly evaporate and become a relic of the past is just part of modern progress, and in many ways it is seen by us as inevitable—but is it truly inevitable, or is it not borne out of the choices we are making?

Ultimately, it feels like a choice based on the cultures' and peoples' value to us, specifically an economic one. That may sound inhumane or compassionless, but it is a hard truth of our modern society. There are countless examples of large-scale industrial projects in rainforests across the world that prove this to be true, where the ancestral lands and homes of many native peoples are flooded, buried, contaminated, or undermined. There is significant importance in highlighting the plight and injustice of Indigenous people and their land through the media and humanitarian organizations and campaigns; of even greater importance is to recognize the wisdom, the perspectives, and the practices of Indigenous people as a foundation for, and of critical value to, the global knowledge economy and survival of humanity, ceasing to measure their significance as a mere object for tourism or as craftspeople creating colorful bags. Only then can we understand their value and the enormity of what true value really means, leading us to find which answers to the numerous questions about the critical state of our times the Kogi can guide us toward.

THEN THE WATERS WILL BE CLEAN AGAIN

"How do we live a good life?" Mama José Gabriel often asked me.

While such a question is pertinent for us, we seem more concerned with questions like: "How does something function better? How can we achieve greater results? How do I make it more enjoyable? How do I make it easier and speed up the process?"

While the Mamo's question at first sounded a little banal, it actually encompassed precisely what needs to be addressed if we intend to follow the Kogi's message and begin to comprehend their approach of aligning with the aliveness of all things. Environmentalist Paul Hawken writes: "Business and industry is the only institution that is large enough, pervasive enough and powerful enough to lead humankind out of this mess."[7] Corporations, and the people that form them, are the current driving force of the systems that, for now at least, prop up society and that strive, often without integrity, to deliver innovation, progress, and material well-being, but they also are the catalysts for great inequality, exploitation, destruction, and the malnourishment of our societal spirit. With the business industry being the primary source of financial income for most people, we are still reliant on its current structure to maintain the global status quo, but we are at a crossroads where new possibilities are quietly emerging.

We are currently living in the age of information, of data becoming a vastly more valuable resource than the people it stalks, measures, and charts. The growth of technology is both alarming and astonishing; the speed at which it is evolving is, worryingly, so fast we seem destined to be unable to fully grasp the effect it has, and will have, before it is potentially too late.

The explosion in 2023 of AI technology becoming mainstream, with companies such as OpenAI producing ChatGPT, allowing anyone with basic computer skills to collect, produce, and analyze information and create a blurred line between what is real and fake, all for free and at an unprecedented speed, perpetuates a society unable to distinguish reality from fiction.

In an increasingly fast-paced and volatile world that is ever more prone to interference and subterfuge, new and different qualities and ambitions will emerge. Change must always begin with the individual, and then like a plant that first germinates from a seed, it can grow and sprout branches and then leaves that can have far-reaching effects on everyone and everything that they come in contact with.

Yet currently, individuals are bound to conditions and systems that either allow or forbid such development, or at the very least are strongly incentivized to adhere to specific behavior and codes of conduct. Simultaneously, transformation can never solely be just about the individual or be solely on a systemic level, but must always include the whole to make a true shift.

Futurist John Naisbitt declared: "The greatest breakthroughs of the 21st century will not come from technology, but from an expanded concept of what it means to be human."[8]

This is where Indigenous peoples in general, and the Kogi in particular, can offer the greatest inspiration and value to us. Shifting our perspective of what it truly means to be human begins by honoring our individuality, and in terms of a company, by acknowledging it to be an organism and not a machine. If a company perceives itself as a machine, in the long term usually it can only be run by a great deal of motivation and control, and the humans involved simply become objects that are worn down. On the other hand, when a company functions as an organism, the conditions for it to flourish naturally arise. A machine can be smart and of high technological intelligence, yet an organism is wise, present, perceptible, and always comprises life force.

Life force energy is also the foundation of true balance. Does the bee deplete the flower of its nectar? Or does the flower use the bee for pollination? Neither applies, as this relationship is natural to both of them, and they both benefit and give each other what they need. This reciprocal interaction can be seen as being filled with life, because it is just as innate for the bee to pollinate flowers even while its main focus is to fulfill its own needs.

Therefore, by creating companies and organizations with life force as their foundation and leading principle, we create a healthy and thriving organism, or rather, we simply let it evolve into being. This requires us to let our thinking be guided by life force.

In nature there is no waste or energetic deficit, and as the Kogi would say, the thought of it does not even exist. Every redundancy in nature serves as a resource,[9] as should every surplus in a company. Nature's enormously efficient use of what is available and the fact it does not waste energy, while also offering infinite abundance of life force, make it both beautiful and elegant. For example, when cherries are ripe on a tree, there is suddenly such an abundance that there are more fruits than can be eaten.

Organizations that are structured around "livingness" do not waste energy on unnecessary meetings, prestige, debilitating working hours, or power games. Organizations such as these, however, do not withhold an abundance of energy when investing in employees and their workplace. Such a company or organization is similar to an organism in that it takes something from nature, uses it, but then gives something back in return. Companies that work together with nature, in the sense of the One Thought of the Kogi, will not only survive, but thrive. The Kogi would say that the earth supports them to do so. If these organizations are aligned with nature and are based on the principles of life force, they will automatically only make decisions that do not harm nature, because it is their intrinsic nature to think this way.

In nature, strength is often achieved by making structures more flexible. We can observe this in a leaf; its structure exists because of the water that flows through it. A leaf does not buckle in a storm, but a large solid tree does. Geese fly in a V formation on their migrations; it has been found that this saves a lot of energy, because each goose produces lift for those behind. Because of this they can cover over 70 percent more ground; each goose is immediately aware when it leaves the formation—it is much more strenuous to fly alone. When the leading goose gets tired, it lets itself fall back and integrates into the V formation and another goose takes the place at the front. How do the geese know where to fly? It is their intrinsic wisdom and is always within them, their flight path is who they innately are.

REESTABLISH OUR PLACE

The Kogi Mamos and Sakas are the vision-holders and guides of their people. Their work not only strengthens their people but also enhances all of nature. If nature is strong, the Kogi are strong, and vice versa.

What can we learn from them that applies to our own development of cooperation, innovation, and value creation?

Many corporate mission statements from companies advocate for the removal of hierarchal leadership, for greater shared responsibility, more sustainable approaches, better communication, greater resilience and reliability, and so on. For the Kogi, however, these are not just empty words or ambitions, but are lived practices. A natural leader, just like a Mamo or a Saka, is inspired and inspiring, and as one of the central figures of his community, is focused more on processes than achievements.

Many Indigenous peoples have never lost this style of guidance. The Kogi are currently already living what many individuals, and companies, aim to achieve in the future. They are decentralized, nonlinear, and therefore as deeply interconnected as life itself, and still have a clear order and governance structure. Once we again dare to see Indigenous knowledge and ways of life as part of a "knowledge economy," we can learn a lot.

Nature's proficiency in solving organizational challenges is immense, and Indigenous peoples have been observing and understanding nature for thousands of years.

BECOMING INDIGENOUS TO LIFE

In modern culture we have forgotten the necessity for "rites of passage," which are used traditionally as marking a transition from one stage of life to another, commonly from childhood to adulthood. It is a symbolic journey that helps individuals discover their place in the world, and their role in the community. This means recognizing that we are part of something greater than ourselves. It means embracing our interconnectedness and working toward the well-being of the whole. This requires the willingness to learn from the natural world.

Life, in itself, is a conglomeration of regenerative communities of differing proportions; from the smallest microorganisms to the largest ecosystems, life is constantly renewing itself. By understanding this innate regenerative nature of life, we can learn to live in harmony with it and start to reinhabit life.

One way of achieving this is to become experts at being aware of our actions and our impact on the natural flow of life. We must develop a strong sense of awareness and a deep, intuitive perception. By paying attention to the rhythms of nature, we can also align ourselves with its flow and find our place within it. If, however, things become a struggle and out of flow, we must reexamine ourselves, our motives, and our direction. This may mean taking a step back, reflecting on our values and goals, and seeking guidance from the natural world.

Some individual practices of connecting with nature can be as simple as taking a walk in the forest, gardening, or observing the stars. By developing a regular practice of connecting with the natural world, we can deepen our understanding of our place within it and find our unique role and purpose within the greater web of life.

Whatever our role may be, it is important to recognize that we are not alone in this work. We are part of a larger community of individuals and organizations who are also striving to live in harmony with the earth. By connecting with others who share our values and goals, we can amplify our impact and create positive change on a larger scale.

Becoming Indigenous to life is an ongoing process. It requires continual learning, growth, and adaptation as we navigate the complex challenges of our time. But by embracing our place within the greater web of life, we can find meaning, purpose, and fulfillment while contributing to a healthier, more sustainable future for all.

TOGETHER WITH THE EARTH

From writing this book, and from spending time with the Kogi, I have realized that the most crucial and urgent point is this: our core challenge as a collective humanity, and as individuals, is how to become true counterparts for the earth, instead of misusing our undeniably unique place within creation as dominators, so we can become guardians again.

Not only do we need the earth to exist and thrive for our own survival and existence, but the earth also needs us to thrive—the Mamos continually assured me of this potentially contentious point. The earth may exist without humans, but the earth will not thrive. There is a fundamental reason why humans exist on earth; we are not simply parasites or an accident. We have great capacity for innovation, ingenuity, and compassion—and we have the physicality, resilience, and innate resources within us to offer new paradigms of change. Becoming a guardian is a long and constant journey of learning, of adapting, of engaging and listening—but pivotally, connecting.

We must remember new ways of connection. Yes, remember, because it is already within us. With this openness and receptivity, we can perhaps begin to comprehend the brilliance of the world we live in and be emboldened to protect and cherish it over all else, just as we would our own children, or our mother.

By adapting and dedicating ourselves to a life lived in reciprocity, we establish a world that allows nature to govern, to control, to inspire, and to lead—we create a world that is foundationally, at its core, a planet that is self-replenishing and balanced, and a global civilization that is Indigenous to life.

This evolution of consciousness is a reciprocal process that can take place in any of our relationships: with ourselves, with others, and with the earth. Ultimately, we are not alone in this. The Mamos always say: If we help the earth again, the earth will start to help us too.

Acknowledgments

The Kogi people: First, I would like to express my deep gratitude toward the Indigenous Kogi people of Colombia for sharing their wisdom with me. Without them, my life would be very different right now and this book would not have been possible.

Arregoces Coronado-Zarabata: I am especially grateful for my dear Kogi friend Arregoces. With his wonderful humor, he accompanied me on my many journeys into the sacred mountains of the Sierra Nevada de Santa Marta. He translated most of the original wisdom from the different Mamos and keeps making fun of me in the wittiest and kindest way. This friendship really means a lot to me. I also thank his family: Rosa, Ana Maria, and Juan David for always receiving me well and for coping with Arregoces's long absences when we travel together all over the world.

Fabienne Balmer: A big thanks goes to Fabienne. She undertook the enormous task of translating the book into

English at a time when we didn't even have a publisher, or any certainty that the book would be published at all, and she did so solely for the message of the Kogi to be brought into the world.

Jackson Gore: This book would not have been possible without the amazing work of Jackson. In long hours of challenging work, he not only edited the English translation, but adapted the book with his great love for detail, contributing the literary glue that makes this book the read that it is right now. Thank you very much for your effort to bring the message of the Kogi to the English-speaking world.

Notes

Before the Journey
1. The *k* is pronounced deep in the throat, similar to the Arabic letter *qaf*.
2. The *nuhué* is a ceremonial house of the Kogi men.
3. Sezhankua or Serankua is the creator.
4. Lucas Buchholz recorded and transcribed most of the texts quoted in this book. The rest was kindly provided by Arregoces Coronado-Zarabata.
5. *Esuamas* are places of power and knowledge.

Chapter One: What Did You See Today?
1. The Kogi do not see thinking as purely a mental act; it also includes feelings and sensations.
2. Jaba Sé is the mother of all existence. Sezhankua (or Serankua) is the creator of the earth. Aluna is the world of thoughts and the intelligence of nature.
3. The Kogi call all non-Indigenous people Younger Brothers and Sisters.
4. A *pagamento* is an offering to restore balance that is often given to the earth.
5. Mamos are the wise elders of the Kogi. Mama is used to address a specific Mamo, denoting a familiarity, while Mamo is used when speaking generally.
6. See Wade Davis, "Dreams from Endangered Cultures," Filmed February 2003, TED Talk, https://ted.com/talks/wade_davis_dreams_from_endangered_cultures?language=rm.

Chapter Two: What Thoughts Have You Been Thinking?

1. On the hat of a Mamo, different lines represent different climate zones.
2. For more information on the cycles of water, see *Aluna*, directed by Alan Ereira (Sunstone Films, 2012).
3. *Linea Negra* means "black line," which is the boundary of the original territory of the Kogi at the foot of the Sierra Nevada de Santa Marta. Many sacred places are located on this line.
4. Jate Teikú is a figure in Kogi mythology.
5. So it is not surprising that the Kogi also consider the wind to be alive.
6. Gerardo Reichel-Dolmatoff, *Los kogi, una tribu de la Sierra Nevada de Santa Marta (The Kogi. A Tribe of the Sierra Nevada de Santa Marta, Colombia)*, vol. 2 (Bogotá: Editorial Iqueima, 1951), 94.
7. *Alúnayiwási* is the confession of the Kogi, in which everything positive and everything negative that was said, thought, or done is revealed.

Chapter Three: Do You Know the Stories Well?

1. The Organización Gonawindúa Tayrona are political representatives of the Kogi, Wiwa, and Arhuaco with offices in Santa Marta and Valledupar.
2. All place-names are changed to protect the Kogi. Altamira refers to a village in the south of the Sierra that is the center of what is left of the Kankuamo culture.
3. "She's right, she's right about certain things, because you'll blame me for being a womanizer," from "El Condor Herido."
4. Translation: "I'll take the female."
5. See also by James Lovelock, *Gaia: A New Look at Life on Earth* (Oxford: Oxford Univ. Press, 1979); *Ages of Gaia* (Oxford: Oxford Univ. Press, 1995); *The Revenge of Gaia: Why the Earth Is Fighting Back* (New York: Allen Lane, 2006). As well as Lynn Margulis, *Symbiotic Planet: A New Look at Evolution* (New York: Basic Books, 1999).
6. A narrative is the basic story (that which is told) and thus a view of a situation, an event, an idea, and even of the world as a whole.
7. See Genesis 1:28.
8. Gerardo Reichel-Dolmatoff, "The Great Mother and the Kogi Universe: A Concise Overview," *Journal of Latin American Lore* 13, no. 1 (Summer 1987): 112.
9. Nuanase is a negative figure in the mythology of the Kogi.
10. *Kwalama* are cycles that sustain life, food security, and fertility ceremonies for children and seeds.
11. Subir Bhaumik, "Tsunami Folklore 'Saved Islanders,'" *BBC News*,

January 20, 2005, http://news.bbc.co.uk/2/hi/south_asia/4181855
.stm; Roger Highfield, "Did They Sense the Tsunami?," *Telegraph*, January 8, 2005, https://www.telegraph.co.uk/technology/3337851/Did
-they-sense-the-tsunami.html; "Tsunami, Ten Years On: The Sea Nomads Who Survived the Devastation," *Guardian*, December 10, 2014,
https://www.theguardian.com/global-development/2014/dec/10
/indian-ocean-tsunami-moken-sea-nomads-thailand.

Chapter Four:
What Did You Plant? What Did You Harvest?

1. *Mamatishé* is Damana, not Kággaba. Damana is a dialect of Kággaba, similar to the Wiwa language.
2. Mamayuisa is Damana, not Kággaba.
3. Éric Julien, *Kogis—Le message des derniers hommes* (Paris: Albin Michel, 2004), 255.
4. Kwivi, in addition to being a Mamo student, is also a name for a Mamo or Saka school.
5. When the Kogi speak Spanish, they use the word *pagamento* as a translation of their word *zabiji*. *Pagamento* means payment, and in this context means offering in the sense of establishing balance, just like we pay money for a product purchased in a store. If this payment is not made, one has liabilities that still have to be settled. The Kogi even use the word *deuda*, meaning debts in Spanish. This term relies on the concept, both in our culture and that of the Kogi's, that came from the Christian church. The Kogi unfortunately suffered many attempts at the hands of the Christian missionaries to force their religion and knowledge on them. Therefore, I have decided to use the terms *imbalances* or *liabilities*, because what the Kogi mean has nothing to do with our concept of debts or guilt.
6. Mama José Gabriel refers to a current situation at that time: Wayúu Indians starved and died of thirst on the Guajira peninsula due to a total lack of water.
7. Comment by Arregoces Coronado-Zarabata: "What I don't know is how the Mamos know that when it's night here, it's day somewhere else. Some have never left their mountain, they have never even been here in Santa Marta. They see these things in the darkness within themselves. The Mamos have vast ancient knowledge. Mama Shibulata said that our world is one united world. At the very beginning of the world there was also only one continent, but then it split and the sea flowed between them. When I was in Europe recently, they confirmed exactly this."
8. Juan Carlos Mamatacan, who was translating for his grandfather, noted

that it's like Google on the internet, where you ask a question and get one or several answers.

Chapter Five:
How Will We Pass the Knowledge On to Our Children?

1. Panela is a product similar to sugar, obtained by boiling the sugar molasses and then pouring it into cube-shaped molds and drying it.
2. Kággaba is the language of the Kogi and the name they call themselves, which means "human." However, since they are internationally known as Kogi, that is the name used in this book.
3. Salt creates emotional residue in the body, and this disturbs the flow of life force in our thoughts.
4. Not to be confused with right and wrong.

Chapter Six:
What Do I Have to Offer the World?

1. Gerardo Reichel-Dolmatoff, "Training for the Priesthood Among the Kogi of Colombia," in *Enculturation in America, An Anthology*, ed. Johannes Wilbert (Los Angeles: Univ. of California, Latin American Center, 1976).
2. Mamatungwi is in Kággaba Mukuakukui.
3. Sé Shizha is the Linea Negra—the territory of the Kogi.

Chapter Seven:
What Does It Mean to the World That I Am Here?

1. Coca is a sacred plant for the Kogi, and for many other indigenous people throughout South America. It is a natural and highly nutritious stimulant and is far distant from the chemical derivative substance cocaine.

Chapter Eight:
How Will We Make Sure That Everyone Is Well?

1. Éric Julien, *Kogis—Le message des derniers hommes* (Paris: Albin Michel, 2004), 126.
2. *Ruamashka* is a term from the Wiwa language.

Chapter Nine:
How Will We Live Another Eighty Thousand Years?

1. For the following examples in this subchapter, see Janine Benyus's two lectures at TED: "Biomimicry's Surprising Lessons from Na-

ture's Engineers," filmed February 2005, TED Talk, https://www
.ted.com/talks/janine_benyus_biomimicry_s_surprising_lessons
_from_nature_s_engineers?language=en; and "Biomimicry in Action," filmed July 2009, TED Talk, https://www.ted.com/talks
/janine_benyus_biomimicry_in_action?language=en.

2. See *Kiss the Ground,* a film produced and directed by Josh Tickell and
    Rebecca Tickell (Big Picture Ranch, 2020), for more details.

3. Janine Benyus, "12 Sustainable Design Ideas from Nature," filmed May
    17, 2007, TED Talk, https://www.youtube.com/watch?v=n77BfxnVIyc.

4. Janine Benyus, *Biomimicry: Innovation Inspired by Nature* (New York:
    William Morrow, 1997).

5. David Suzuki, *Sacred Balance: Rediscovering Our Place in Nature,* 3rd ed.
    (Vancouver: Greyston Books, 2007).

6. Benyus, *Biomimicry;* Benyus, "Biomimicry in Action."

7. Paul Hawken, *The Ecology of Commerce: A Declaration of Sustainability*
    (New York: Harper Business, 2010).

8. John Naisbitt, *Megatrends: Ten New Directions Transforming Our Lives*
    (New York: Grand Central, 1988).

9. A redundancy is the presence of double and seemingly unneeded re-
    sources.

About the Author

Lucas Buchholz, born in 1989, is focused on transforming the consciousness of society and economical institutions toward regenerative models. He is a consultant for innovative regenerative projects and businesses, an author of two books, and an inspirational speaker who acts as a world bridge between cultures, highlighting how our modern societies can learn from ancient and indigenous ways of life and knowledge. His life, and work, is guided by one fundamental question: How do we become indigenous to earth again? One key to this is the wisdom of the Kogi people, who he has been in contact with for many years. He founded an organization that supports the ongoing growth of an educational academy as well as a film project. Lucas lived with the Kogi people in Colombia. For more information, visit lucasbuchholz.com/en.